Irrigation Pumping Plants

by
Blaine Hanson,
UC Irrigation and Drainage Specialist

Division of Agriculture and Natural Resources Publication 3377
University of California Irrigation Program
University of California, Davis

Funded by the California Energy Commission and the U.S. Department of Agriculture Water Quality Initiative

Irrigation Pumping Plants

by
Blaine Hanson,
UC Irrigation and Drainage Specialist

University of California Irrigation Program
University of California, Davis
Rev 2000

Funded by the California Energy Commission and the U.S. Department of Agriculture Water Quality Initiative

Water Management Series publication number 3377

ORDERING INFORMATION:

Copies of this publication can be ordered from:

1-800-994-8849, or
www.anrcatalog.ucdavis.edu

Other publications in this Water Management Handbook Series include:

Agricultural Salinity and Drainage (publication number 3375)
Surge Irrigation (publication number 3380)
Micro-irrigation for Trees and Vines (publication number 3378)
Drip Irrigation for Row Crops (publication number 3376)
Surface Irrigation (publication number 3379)
Scheduling Irrigations: When and How Much Water to Apply (publication number 3396)

Designed and edited by Anne Jackson
Cover art by Ellen Bailey Guttadauro

Published in the United States of America by the Department of Land, Air and Water Resources, University of California, Davis, California 95616.

500-rep-9/05-BH

Contents

Improving Performance (continued)

Preface

Irrigation Pumping Plants is one of a series of water management handbooks developed by the University of California Irrigation Program. Funding for this project was provided by the California Energy Commission and the U.S. Department of Agriculture Water Quality Initiative. The author wishes to acknowledge the assistance provided by Tony Wong, Ricardo Amon and Charles Brush of the California Energy Commission and the contribution of Anne Jackson in the editing and design of the handbook.

Introduction

Managing energy efficiently — obtaining maximum output for every energy dollar spent — is a principal objective in operating an irrigation pumping plant. The more efficient the pumping plant, the more revenue dollars returned per dollar spent on pumping. But for a pumping plant to operate at maximum efficiency, the plant operator must be thoroughly knowledgeable about how pumping plants work.

This manual has been developed to answer the questions that most frequently plague growers about irrigation pumping plant operation. Organized as a series of short chapters on selected topics and prepared in a semi-technical format, the manual is grounded in the author's own field experience and in evaluations of numerous actual pumping plants. At the heart of the book (pages 79-87) is a trouble-shooting guide, which sets out suggested remedies for the most common problems arising in pumping plant operation. Pages 105-110 present case studies — evaluations of real-world pumping plant problems along with recommended corrective action.

Questions and comments about the manual should be directed to the author, Blaine Hanson, Department of Land, Air and Water Resources, University of California, Davis, CA 95616-8628, (530) 752-1130.

Figures

Figures *(continued)*

Tables

Tables *(continued)*

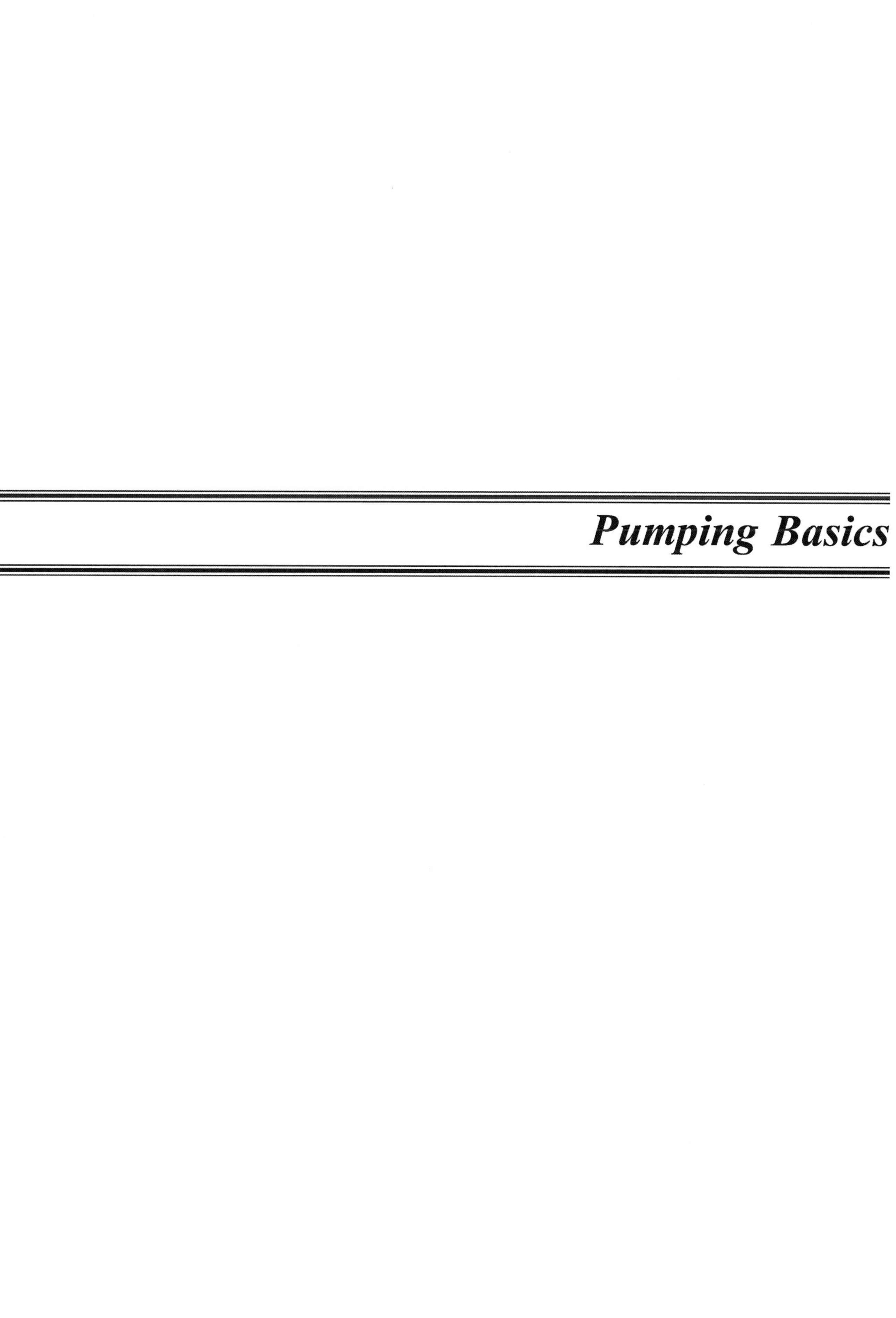

Pumping Basics

How a Pump Works

The types of pumps used most often in irrigation are deep-well turbines, centrifugal pumps, and axial flow or propeller pumps. A pump is made up of a housing or bowl, impellers, and a shaft. The housing or bowl contains the impeller and guides the flow of water from the impeller to the discharge point. The shaft connects the impeller to the electric motor or engine, while the impeller transfers the energy developed by the motor or engine to the water.

In a *centrifugal pump* — which is also referred to as a *booster pump* and which is used to pump surface water into an irrigation system and to pressurize irrigation *systems* — the housing is spiral-shaped, with the diameter of the flow passage increasing toward the discharge point. This type of housing is called a *volute.*

In *deep-well turbine pumps* — which are installed in wells and used to pump groundwater — the housing contains diffuser vanes that channel the water from the impeller to the pump discharge.

Impellers transfer energy from the motor or engine to the water either by centrifugal force or by lifting action. They may be enclosed, semi-open, or open. The vanes of enclosed impellers are enclosed between two plates or shrouds. In semi-open impellers, the shroud is only at the top of the impeller. Semi-open impellers are frequently used in deep-well turbine pumps. Open impellers, which do not include a shroud, are not often used for irrigation, but are used to pump water containing large amounts of solid material.

When the impeller of a centrifugal pump rotates, centrifugal forces develop in the water inside the impeller, which cause water to flow toward the impeller's outer edge. This flow, in turn, results in water flowing into the center or eye of the impeller if the pump is primed.

The output of an impeller depends on the impeller diameter and width and on the rate of rotation. The pressure developed by the impeller depends on the velocity of the impeller's outer edge, which is determined by the impeller diameter and rate of rotation. The impeller flowrate or capacity is determined by the impeller width — at a given rotation rate, the larger the diameter, the greater the pressure, and the greater the width, the higher the flowrate or capacity.

Impellers that pump water only through centrifugal force are classified as *radial flow impellers.* Water flows through these impellers in a direction perpendicular to the shaft. It is characteristic of radial flow impellers that the horsepower demand of the impeller increases as the pump capacity increases.

Axial flow impellers transfer energy to the water by a lifting force. These impellers cannot provide pressure for irrigation systems, but are used in propeller pumps and are capable of delivering very high flowrates at a low pressure. The rotation of an axial flow impeller causes water to flow in a direction parallel to the pump shaft. A characteristic of axial flow impellers is that the horsepower demand decreases as the pump capacity increases.

Mixed flow impellers are often used in irrigation pumping plants, particularly in deep-well turbine pumps. Mixed flow impellers use both centrifugal and lifting forces. A characteristic of these impellers is that increasing the capacity may change the horsepower demand only slightly.

The output of the pump must be sufficient to satisfy the requirements of the irrigation system. Where groundwater is used, the pump must provide the energy required for the pumping lift and discharge pressure (see *Figure 1*). Where surface water is used, the pump must supply the desired discharge pressure but must also be able to supply any suction lift needed at the pump intake. The capacity of the pump must be sufficient to meet the crop's water needs.

Terms

The following terms are frequently used in describing pump performance. These terms are illustrated in *Figure 1*.

- Pump capacity - flow rate at the pump discharge point.
- Static water level - groundwater level in the well before the pump is turned on.
- Pumping water level - groundwater level in the well while the pump is operating.
- Drawdown - difference between pumping water level and static water level.
- Discharge pressure head - pressure (psi) at the pump discharge multiplied by 2.31. This is the height of a column of water in feet that will provide the measured pressure at its base.

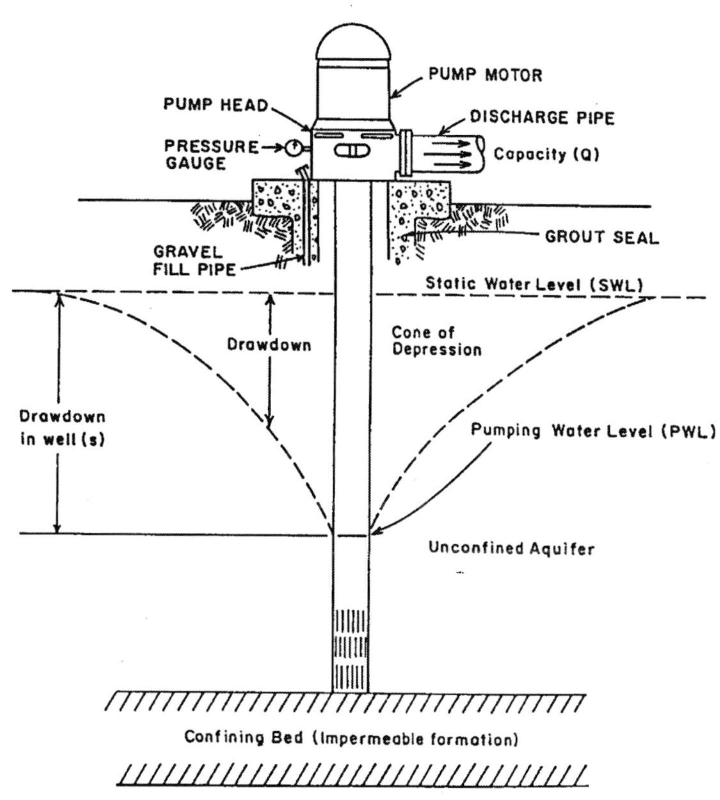

Figure 1. Diagram of a punping plant used for groundwater pumping.

(*Source:* V.H. Scott and J.C. Scalmanini. *Water Wells and Pumps: Their Design, Construction, Operation, and Maintenance.* 1978. University of California Bulletin No. 1889, Division of Agricultural and Natural Resources, University of California.)

Department of Land, Air and Water Resources, University of California, Davis

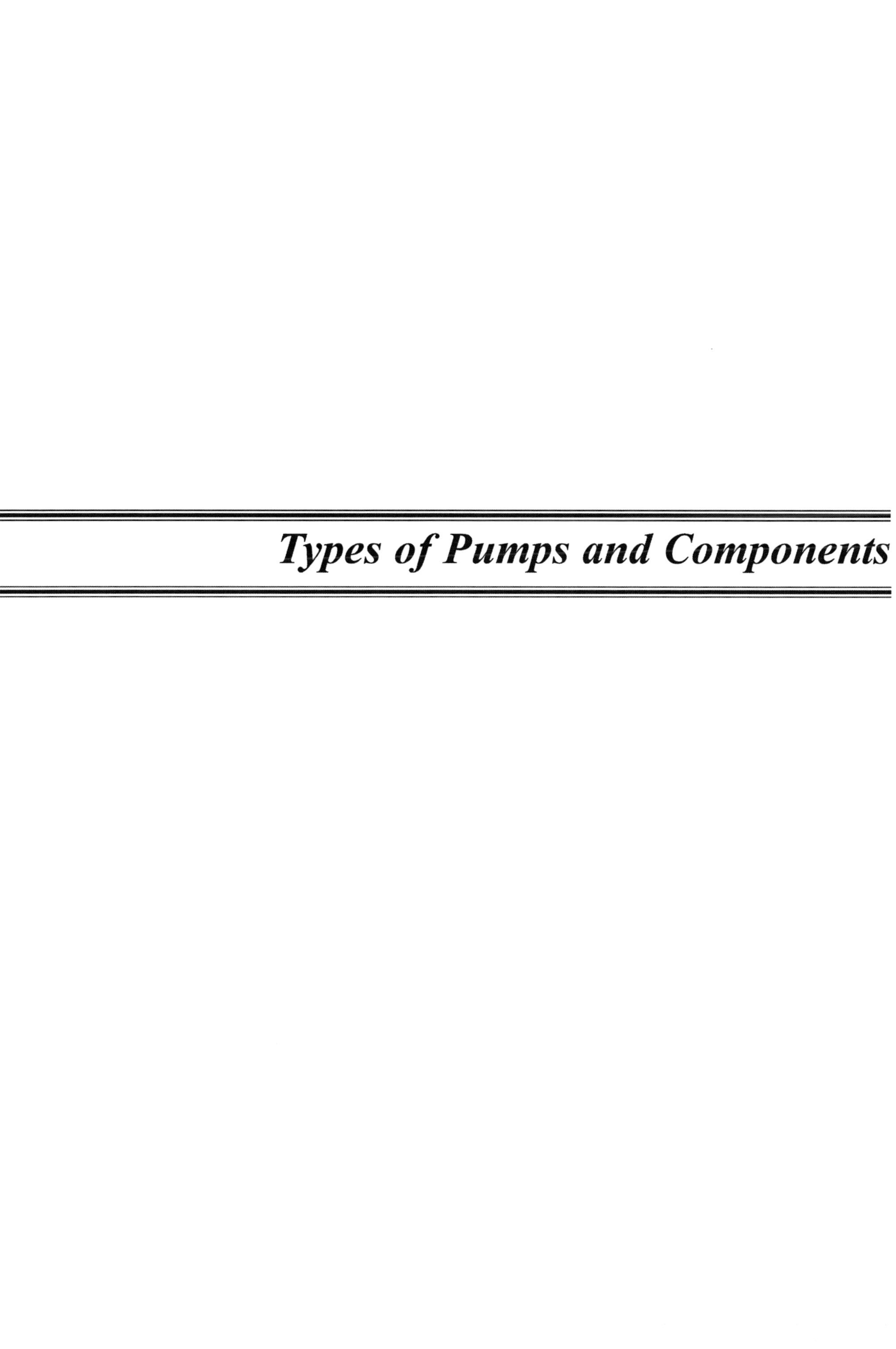

Types of Pumps and Components

Deep-Well Turbines

Deep-well turbines — the usual choice for pumping groundwater — are installed inside the well casing below the water level in the well. The main components are bowls, impellers (located inside the bowls), shaft (connecting the impellers to the power source), and pump head (see *Figure 1*).

Since the impellers and bowls must be submerged inside the well, the diameters of the impeller and bowl are limited by the diameter of the well, which in turn limits the capacity and pressure developed by each impeller. This limitation can be overcome if the bowls are installed in stages, with each stage adding to the output developed by the preceding stages. A five-stage pump will therefore have

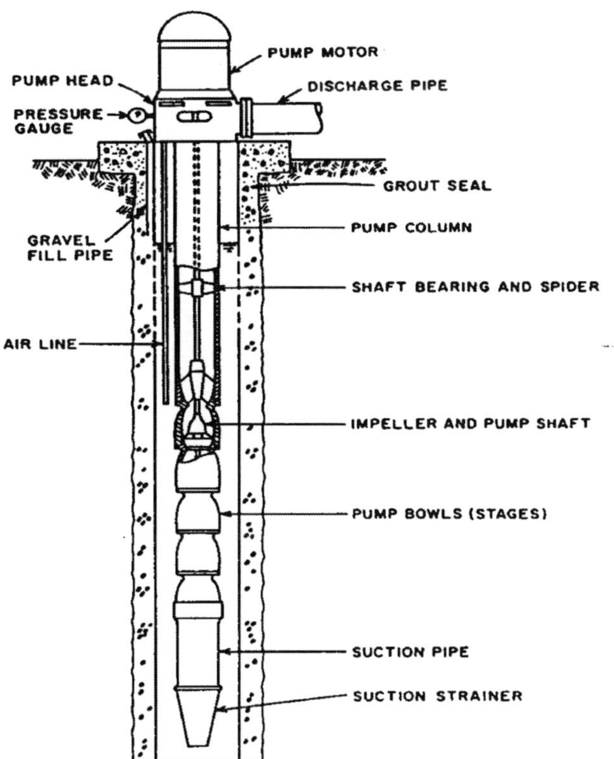

Figure 1. Line-shaft deep-well turbine pump.

(*Source:* V.H. Scott and J.C. Scalmanini. *Water Wells and Pumps: Their Design, Construction, Operation, and Maintenance.* 1978. University of California Bulletin No. 1889, DANR, University of California.)

five bowl assemblies, with the water guided from stage to stage by diffuser vanes built into the bowl assembly. *Figure 1* shows a five-stage pump.

Deep-well turbines use a line-shaft to connect the impellers to an electric motor or engine. Line shafts are supported by bearings — either oil-lubricated or water lubricated — spaced at intervals along the shaft length. Oil-lubricated bearings are housed within a tubing inside the column pipe (the pipe that carries the water to the surface).

The clearance between the bowls and the well casing should be at least one inch. Narrower clearances may complicate installation or, in crooked wells or where gravel becomes lodged between the pump and casing, may make it difficult for the bowls to be removed for repair.

Submersible pumps, which are discussed in the next chapter, are also used for deep-well turbines.

Submersible Pumps

Submersible pumps are deep-well turbines with a waterproof electric motor installed in the well below the pumping level. The motor is attached directly beneath the bowl assembly, and water intake is between the motor and pump. A specially designed motor, slimmer and more compact than a line-shaft motor, is installed in the well. Since both motor and pump are situated inside the well, submersible pumps do not require a line shaft. *Figure 1* shows the components of a submersible pump.

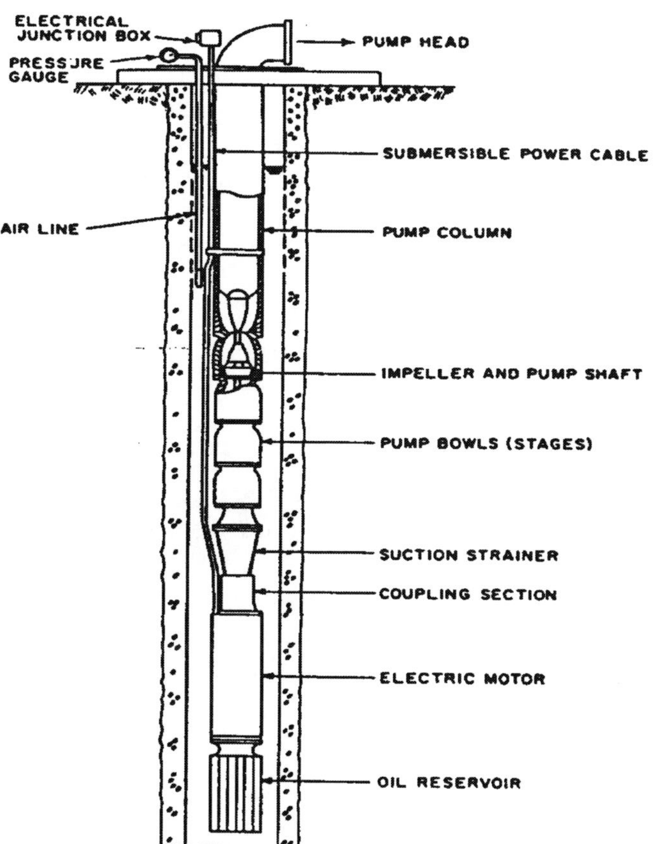

Figure 1. Submersible pump.

(*Source:* V.H. Scott and J.C. Scalmanini. *Water Wells and Pumps: Their Design, Construction, Operation, and Maintenance.* 1978. University of California Bulletin No. 1889, DANR, University of California.)

In submersible pumps it is crucial that enough water circulate around the motor to keep it cool. Since water in the well should flow past the motor casing into the pump intake, submersible motors should be located above the level at which most of the flow enters the well. Water cascading in the well — that is, entering the well at a point above the water level — can carry air down to the motor, preventing the motor from being adequately cooled. A shroud can be installed over the pump intake to force water to flow past the motor.

Advantages

The advantages of submersible pumps are as follows:

- The line shaft normally used on deep-well turbines is eliminated.
- The pumps can be installed in crooked wells. (If in-line shaft deep-well turbine pumps are used in crooked wells, the bearings and line shaft will be subject to excessive wear.)
- Noise levels are reduced.
- The pump motor is protected from vandalism.
- The pump housing can be eliminated.
- Since no line shaft is necessary, submersible pumps may be more economical for very deep installations. Long-line shafts tend to be too heavy for deep installations and are susceptible to mechanical problems.
- Friction generated in oil tubes and line-shafts is eliminated so that pressure loss in the column pipe is reduced.

Disadvantages

Submersible pumps also carry some disadvantages:

- The motor is less accessible for repair compared to a line-shaft system, since to be repaired the pump must be pulled from the well.
- Impellers cannot be adjusted.
- Sand in the water can lock the impellers after the pump is turned off. The pump must be removed from the well to free the impellers. (Sand-locked impellers of a line-shaft pump can be freed by raising and lowering of the impellers from the surface.)
- Submersible motors are less tolerant than line-shaft motors of voltage fluctuations.
- An engine-driven pump cannot be used unless the engine drives a generator.

- Because submersible pumps are "close-coupled", the pump may operate at a very high rpm — close to 3600 rpm. Pumps operating at such high rpms are particularly vulnerable to wear from sand; so if sand is present in the water the pump chosen for a submersible pumping plant should be one that operates at about 1800 rpm.

References

Driscoll, F.G. 1986. Groundwater and wells. Johnson Filtration Systems, Inc., St. Paul, Minnesota.

Halderman, D.A. 1979. Submersible vs. line-shaft vertical turbine pumps. University of Arizona, Tucson, Arizona.

Pacific Gas and Electric Company. 1985. Three-phase service to submersible pumps. San Francisco, California.

Centrifugal Pumps

Centrifugal pumps, usually called *booster pumps,* are placed at the ground surface and are often used for pressurizing irrigation systems and for pumping from surface water sources. Components of these pumps are the volute (impeller housing), the impeller, and the shaft. The volute is spiral-shaped, with the diameter increasing in the direction of the flow. This increase in diameter accommodates the increasing flow rate or capacity as the water nears the pump discharge.

Centrifugal pumps must be primed or filled with water before being operated. When a primed pump is started, the impeller rotation creates a vacuum at the impeller eye, which in turn causes water to flow into the impeller. If the pump is not primed, not enough vacuum will develop and bearings and impeller may be damaged.

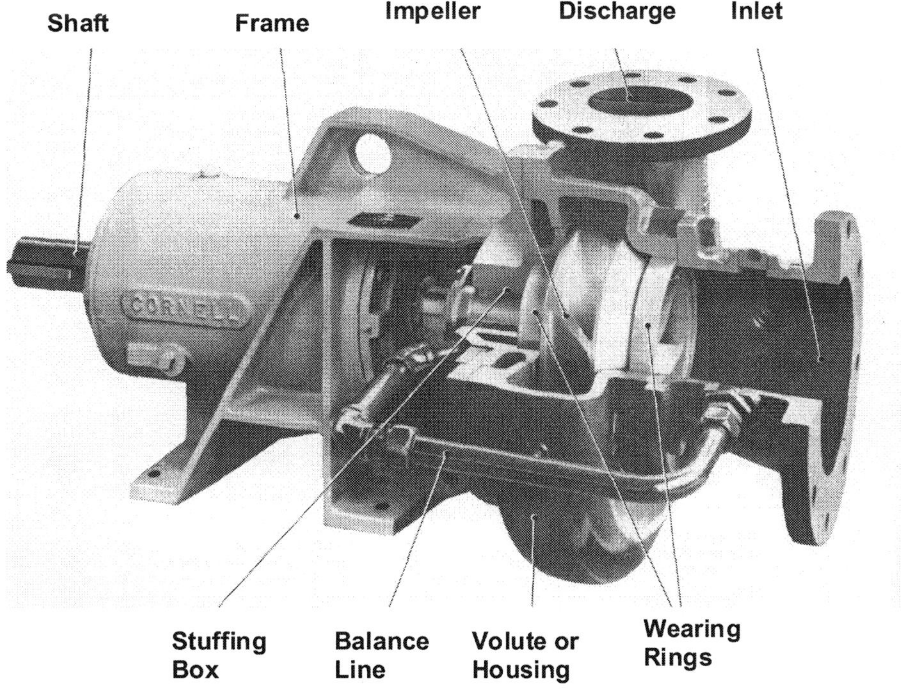

Figure 1. Centrifugal pump.

(Illustration courtesy of Cornell Pump Company, 2323 SE Harvester Drive, Portland, Oregon 97222-7592)

Double Volutes

Some centrifugal pumps employ a double volute design, with a single vane splitting the water flow inside the volute to help neutralize uneven radial forces acting on the impeller, shaft, and bearings.

Wearing Rings

Bronze rings called "wearing rings" are often attached to the pump housing on either side of the impeller in centrifugal pumps — on both the intake side and the stuffing or seal side. To maintain high efficiency, the clearance between the impeller skirts and the wearing rings must be very close. When wear causes the clearances to become too wide, the wearing rings must be replaced. Smaller pumps may not include wearing rings.

Balance Line

Centrifugal pumps sometimes also have a balance line between the stuffing box/seal and the pump intake to help counterbalance unequal forces on the impeller. Balance lines can help reduce pressure along the shaft in the vicinity of the seals and stuffing boxes and also help remove any sand that accumulates near the stuffing or seal.

Suction Lift

Centrifugal pumps installed above the water level will develop a suction lift at the pump intake, causing the pressure at the intake to be less than the pressure of the atmosphere. If the suction lift becomes too great, cavitation will result — a condition in which water vapor bubbles form in the water. These bubbles collapse violently when they reach the discharge side of the impeller. Cavitation can damage the pump and degrade pump performance. To avoid cavitation, suction lifts should not exceed 20 or 25 feet, depending on pump design.

Axial Flow Pumps

Axial flow pumps, which use a propeller to pump water, are appropriate for low-head, high capacity conditions. Components are pump housing, impeller or propeller, and shaft. Pumping is performed solely by a lifting action, with none of the centrifugal action of deep-well turbines and centrifugal pumps. Water is discharged from the propeller parallel to the pump shaft.

The advantage of using axial flow pumps is that flowrates of thousands of gallons per minute can be obtained at very low total heads — less than 30 feet.

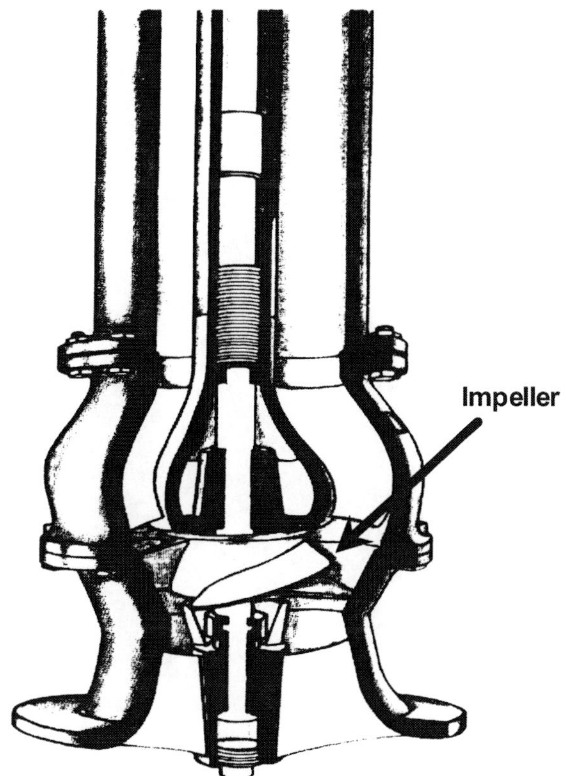

Figure 1. Axial flow or propeller pump.

Impellers

The impeller transfers energy developed by the motor or engine to the water. In deep-well turbines and centrifugal pumps, impellers are rotated at a high rpm, creating centrifugal force. This causes the water to flow toward the rim of the impeller, developing both flow and pressure. In axial-flow pumps, water is pumped solely by a lifting action. Mixed flow impellers — with a combination of both centrifugal and lifting action — are sometimes used in deep-well turbines.

Normally, impellers used in deep-well turbines and in centrifugal pumps are enclosed or semi-open. Enclosed impellers have a disk or shroud on both sides of the vanes (*Figure 1*). These impellers require a close clearance between a wearing ring, located at the inlet of the impeller, and the impeller skirt of bottom extension of the impeller. Semi-open impellers have a shroud only on the top (*Figure 1*). A close clearance (0.003"-0.007") between the bottom of the vanes and the pump housing is necessary for good performance. Semi-open impellers are less susceptible to wear from sand since they can be raised slightly to reduce abrasion. This, however, will decrease pump output and efficiency.

Impellers used for irrigation pumps are normally made of bronze, cast iron, or cast iron coated with porcelain enamel. Enamel cuts down on the amount of pressure lost because of friction inside the impeller. In small pumps, such as those used for domestic water supplies, plastic impellers may be used.

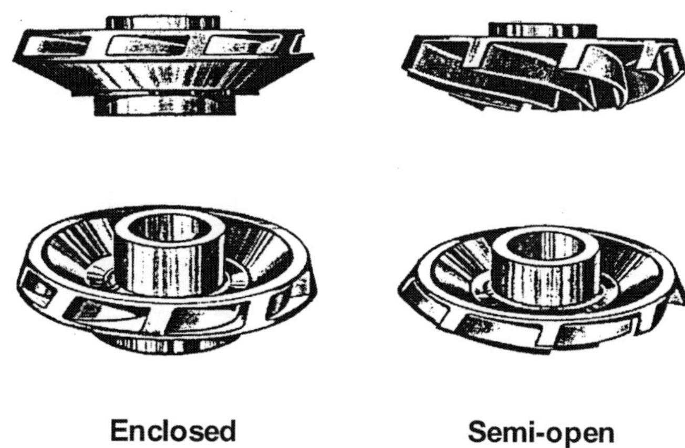

Figure 1. Types of impellers.

(Illustration provided by the Berkeley Pump Company)

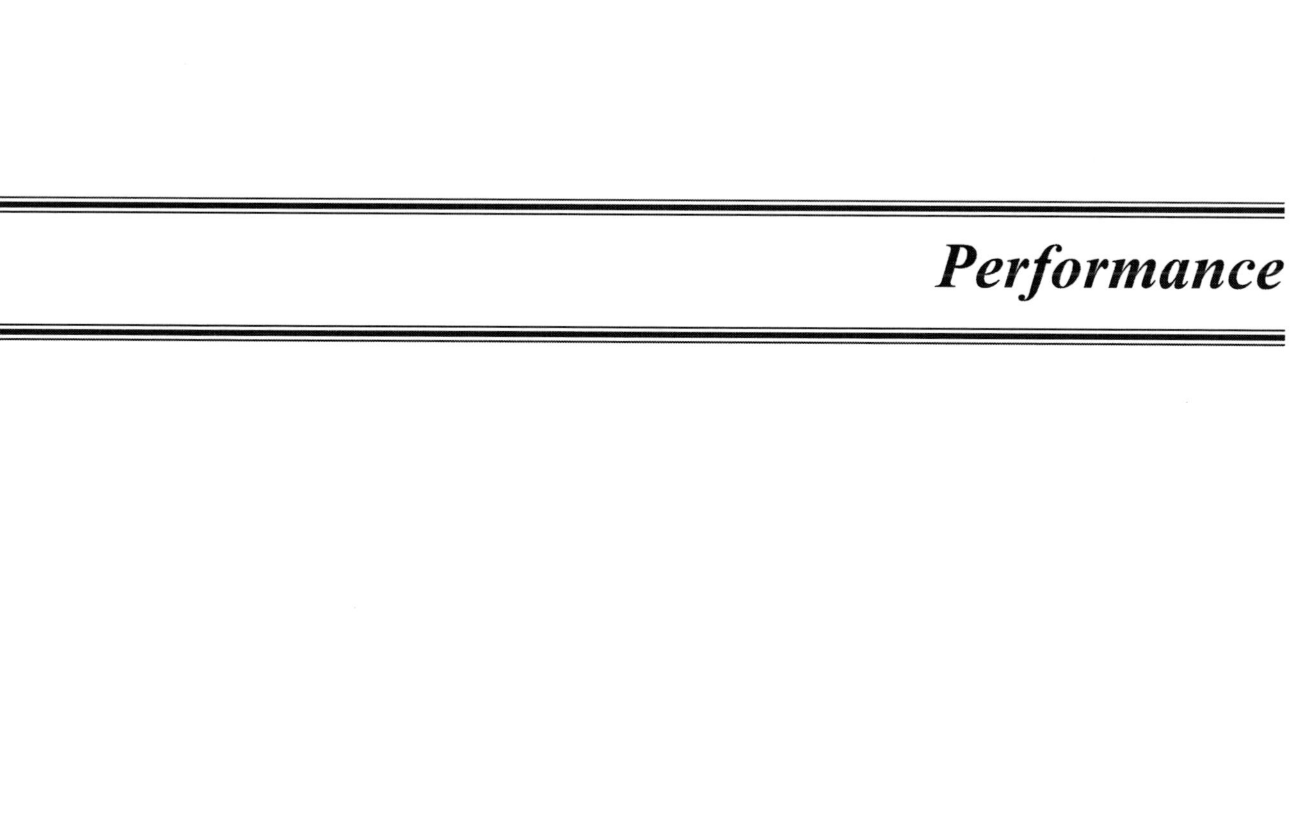

Performance

Deep-Well Turbine and Centrifugal Pump Performance

The performance characteristics of deep-well turbine, submersible, and centrifugal pumps are determined by the design of the particular pump and are defined by performance curves, which describe the amount of head developed by a pump at a particular capacity, the pump efficiency at that capacity, and the horsepower requirements of the pump. Performance curves show which pump will provide the total head and capacity needed by an irrigation system and are also used to determine what size electric motor or engine the pump will require.

Figure 1 shows the total head and capacity performance curve for a typical pump. In a no-flow condition (shut-off), the total head is maximum, or 61 feet for this pump. As the pump capacity increases, the total head generally decreases. At low capacities, most of the pump output consists of total head (normally measured as the sum of the pumping lift and the discharge pressure head), while at high capacities, most of the pump output is in capacity.

Operating the pump at very low capacities should be avoided. If the capacity is too low, overheating of water caused by friction between water and impeller can damage the pump. Also, operating at capacities less than 30 percent and more than 120 percent of the design capacity can increase the radial load on the impeller and cause early failure of bearings.

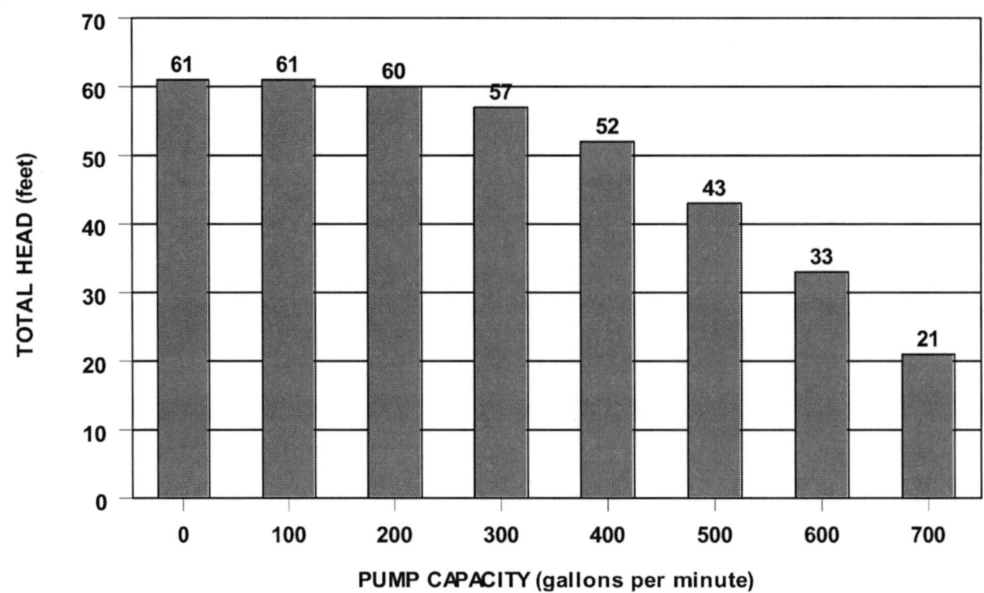

Figure 1. Total head/capacity for typical deep-well turbine and centrifugal pumps.

Figure 2 shows a pump efficiency/capacity performance curve. Efficiency is zero at shutoff (no capacity), increases with capacity to a maximum, and then decreases. The pump selected for the pumping plant should provide the desired head and capacity (that needed by the irrigation system) at the point of maximum efficiency. This will result in the least horsepower demand at that flowrate and total head. The pump that most closely meets this criterion can be identified by comparing the pump efficiency/capacity performance curves of the various pumps available.

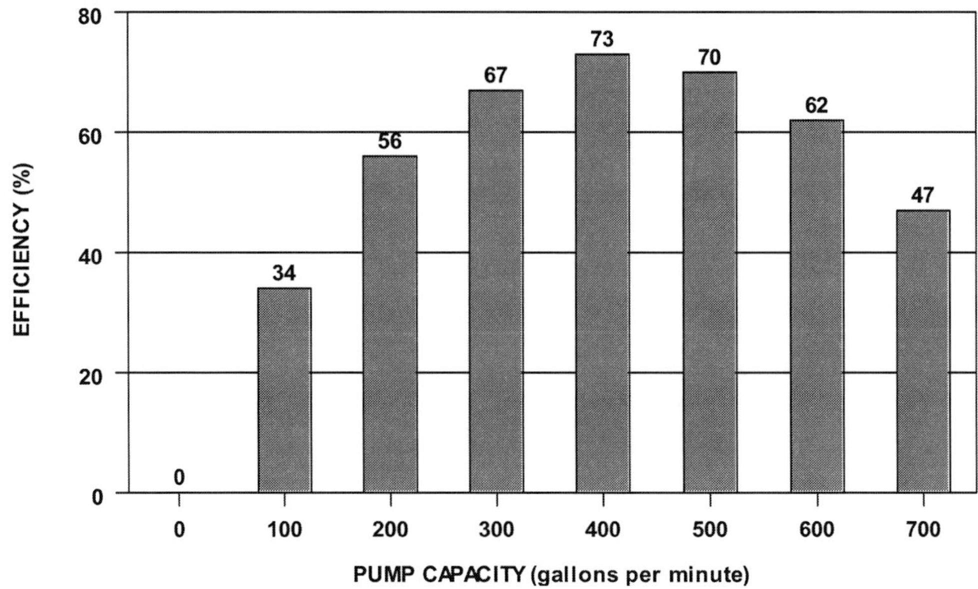

Figure 2. Pump efficiency/capacity performance curve.

Figure 3 shows a brake horsepower/capacity curve. At shutoff, bhp is a minimum 4 bhp, but not zero. Bhp increases as pump capacity increases to a maximum value, which is 8 bhp at 600 gpm. Bhp may decrease slightly thereafter, although in some pumps, bhp may increase at all pump capacities.

Manufacturers provide performance curves for pumps, as well as information on the pump rpm, impeller diameter, and bowl diameter (for deep-well turbines).

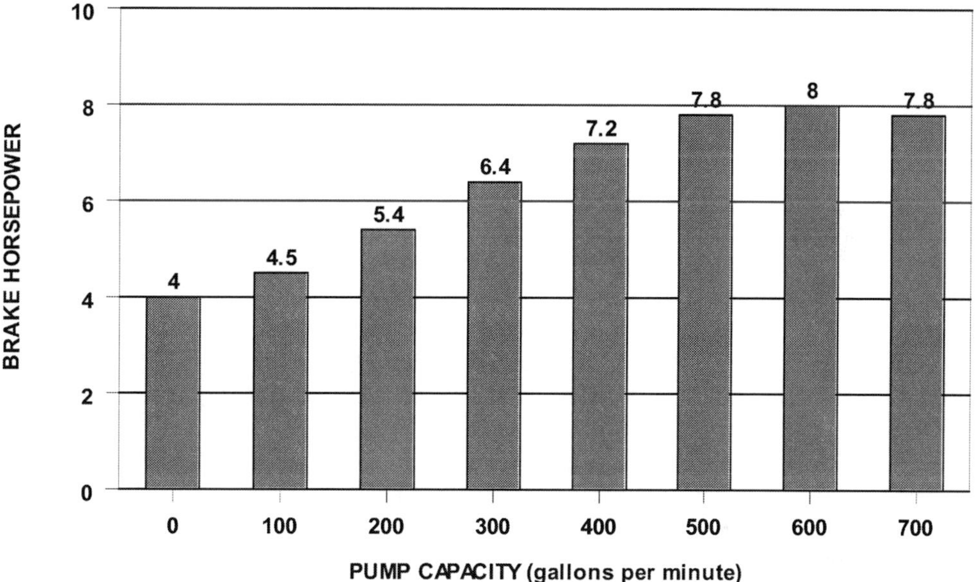

Figure 3. Brake horsepower/capacity performance curve.

Figuring Performance Curves for Multiple-Stage Pumps

The performance curves are only for one-stage pumps, but can be used to develop performance curves for multiple-stage pumps. For example, *Figure 1* shows that in a one-stage pump, 57 feet of head are developed at 300 gpm. The bhp is 6.4. In a three-stage pump, the head developed at 300 gpm will be three times that of a single stage, or 171 feet (*Figure 4*). The bhp will also be tripled, to 19.2 bhp, and the pump efficiency may also change slightly. Usually the manufacturer's curve will show the efficiency adjustments. The performance curve for a multiple-stage pump can therefore be estimated by multiplying the head and bhp of the performance curve by the number of stages for each capacity.

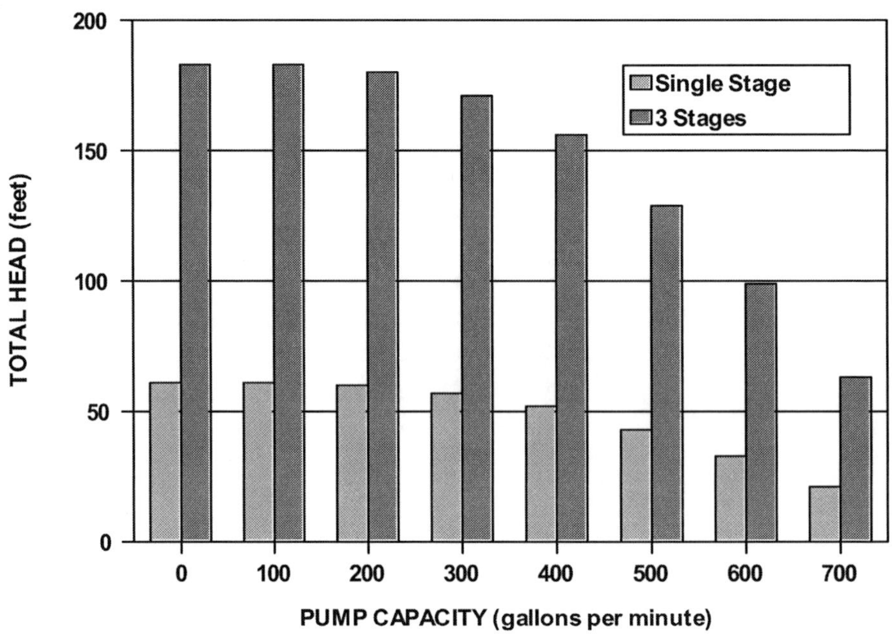

Figure 4. Comparison of performance curves of one-stage and three-stage pumps.

Pump performance curves have a variety of shapes. Some head/capacity curves are steep (large decreases in head occur as capacity increases), while others are flat (small decreases in head occur as capacity increases) (see *Figure 5*). The best curve will depend on changes in pumping conditions. Where pumping levels fluctuate significantly, a steep curve may allow for significant changes in pumping lift with only slight changes in pump capacity. Where sprinkler laterals are added or taken away during an irrigation, a flat curve may help to maintain a relatively constant pressure over a wide range of flowrates.

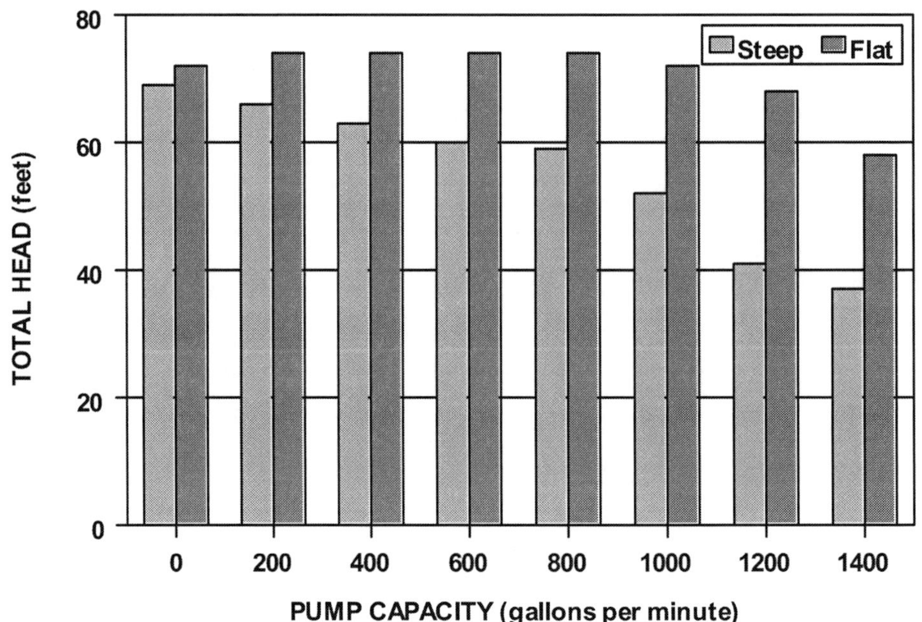

Figure 5. Flat and steep performance curves.

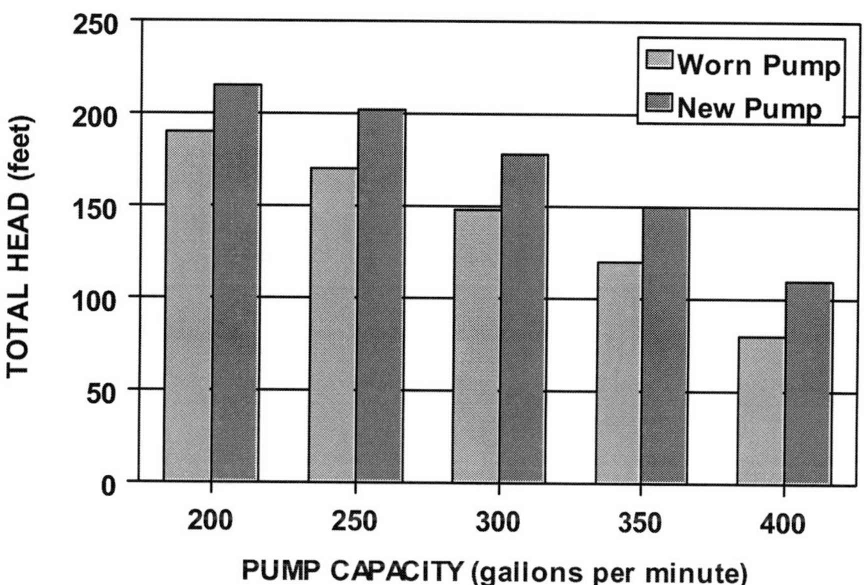

Figure 6. Total head/capacity performance curves for a new and a worn pump.

How Wear Affects Pump Performance

Wear can change the performance characteristics of a pump, resulting in less total head at a given pump capacity and reducing pump or pumping plant efficiency. *Figures 6* and *7* illustrate the effect of wear on pump performance. The principal cause of pump wear is the presence of sand in the water. See page 80 of the *Trouble-Shooting Guide* for suggested measures to counteract wear caused by sand.

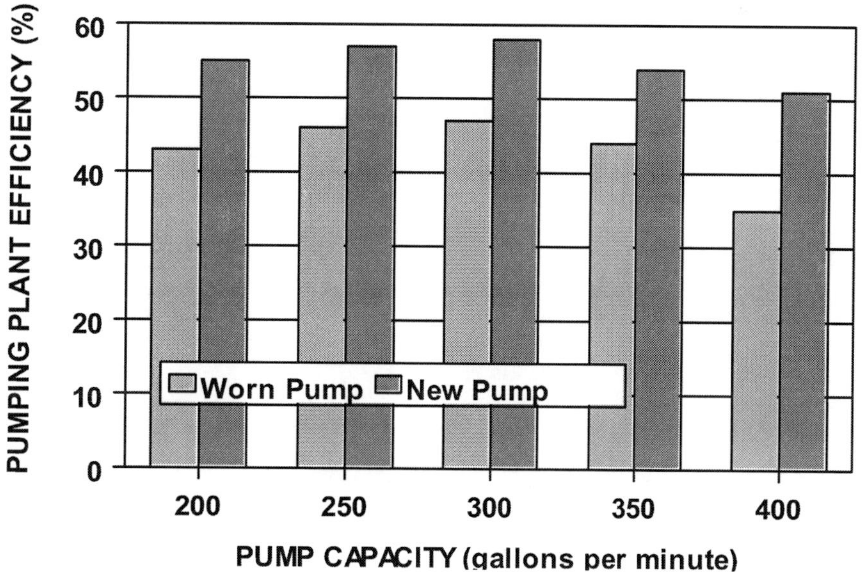

Figure 7. Overall efficiency/capacity performance curves for a new and a worn pump.

While for simplicity's sake, bar graphs are used here to illustrate performance curves, in manufacturer's catalogs performance curves may be presented as line graphs (see *Figure 8*), with one figure showing the head/capacity curve, the efficiency/capacity curve, and the brake horsepower/capacity curve. Curves for various pump modifications, such as different impeller diameters or different pump rpms may also be included in one figure. *Figure 8* illustrates the performance curves for various impeller diameters. Pump efficiency is described here by a series of lines, which form an envelope of efficiency values for the three impeller diameters, instead of by efficiency/capacity curves.

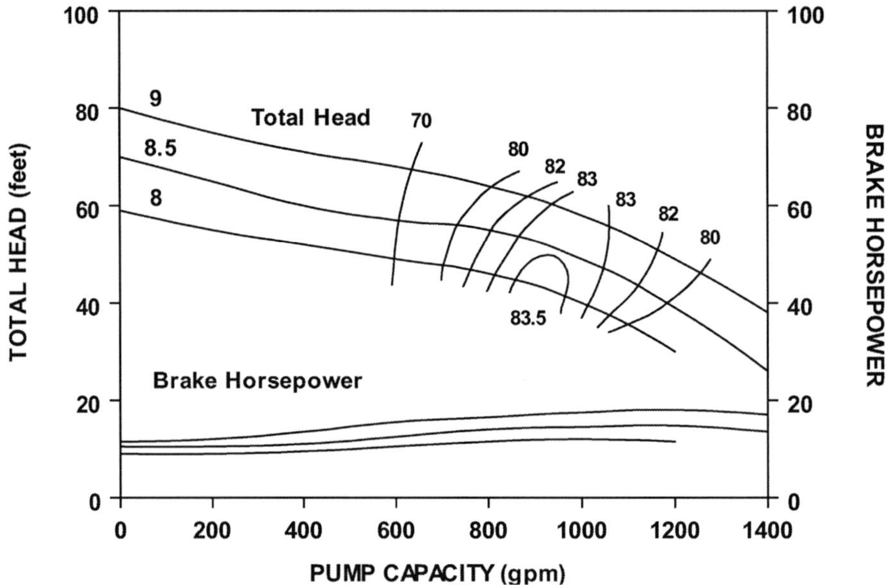

Figure 8. Manufacturer's performance curves.

Axial Flow Pump Performance

The performance characteristics of axial flow pumps differ from those of centrifugal and deep-well turbine pumps. While the total head of an axial flow pump decreases as the pump capacity increases, as in centrifugal and deep-well turbines, the total head/capacity curve of an axial flow pump is very steep (see *Figure 1*). Axial flow pumps can generate only a small amount of suction life compared with centrifugal pumps.

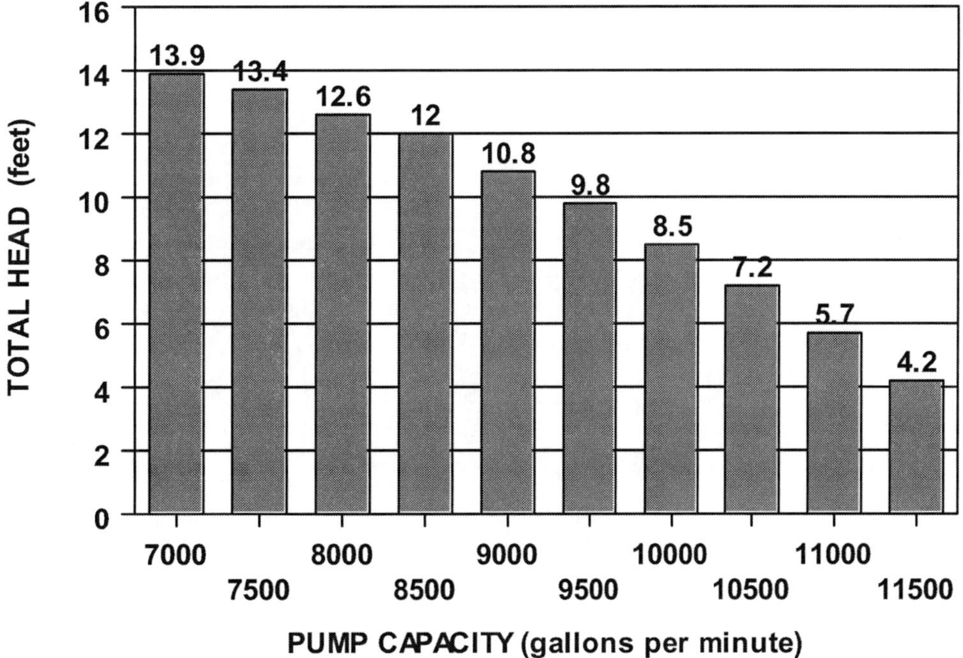

Figure 1. Axial flow pump total head/capacity curve.

The efficiency/capacity curve (*Figure 2*) of an axial flow pump is similar to that of other pump types, but the brake horsepower/capacity curve is very different.

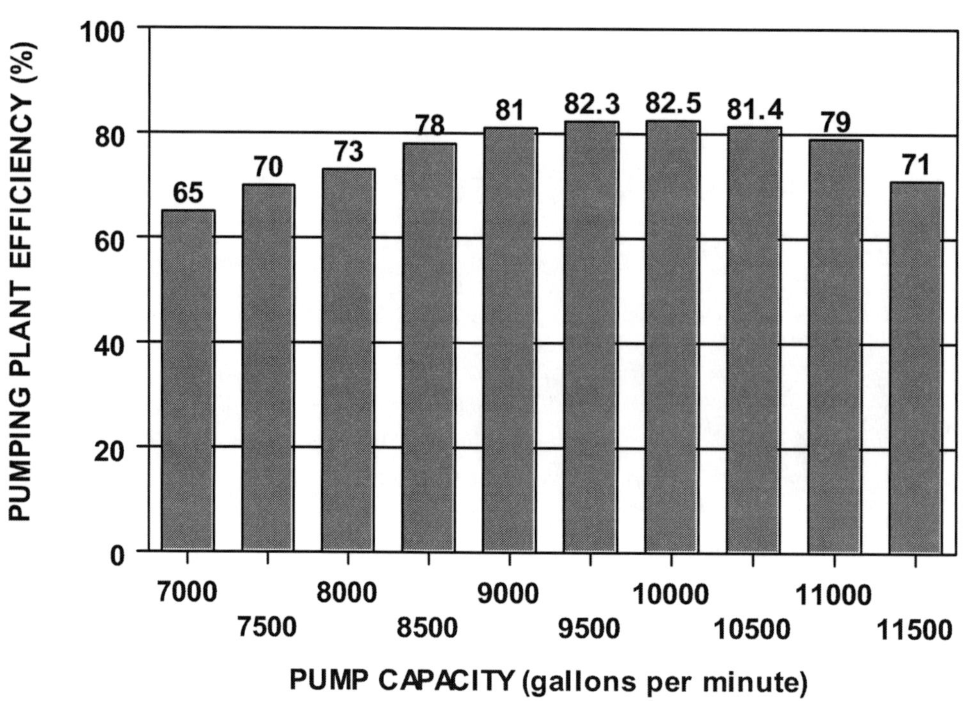

Figure 2. Axial flow pump efficiency/capacity curve.

The horsepower of an axial flow pump reaches maximum at shutoff (no flow condition), with the horsepower demand decreasing as the capacity increases (*Figure 3*). Axial flow pumps, therefore, should not be throttled back, since to do so may overload a motor or engine designed to operate close to the point of maximum efficiency. The rpm of small axial flow pumps can be as much as 1760, while rpms of less than 1000 are common for larger pumps.

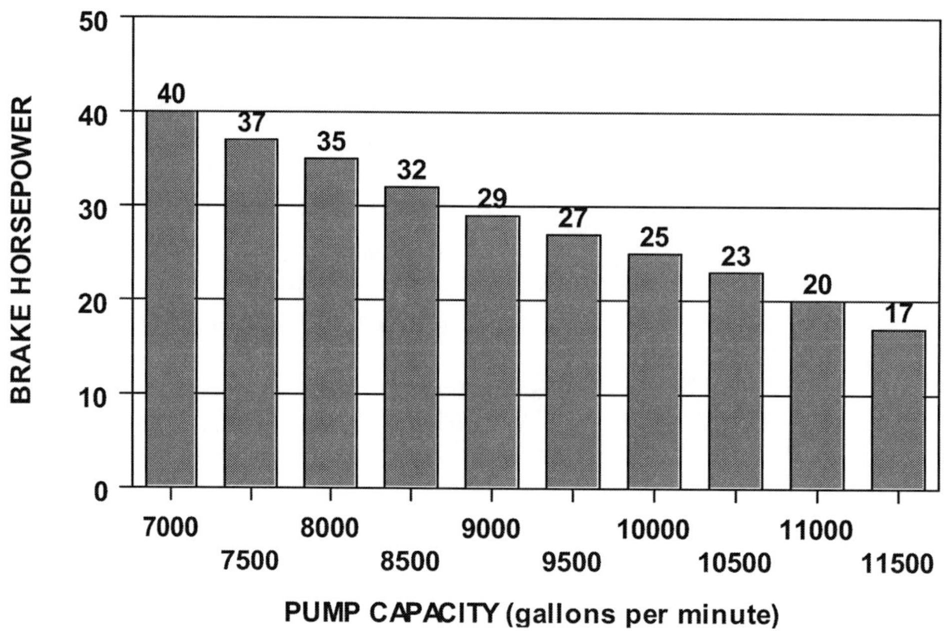

Figure 3. Axial flow pump brake-horsepower/capacity curve.

Submersible Pump Performance

The performance characteristics of submersible pumps are similar to those of line-shaft pumps, since the same bowl assemblies are used, but the question still arises: are submersible pumps more or less efficient than line-shaft pumps? The answer is that the overall efficiency of a submersible pump may be about the same as that of a line-shaft pump.

Contributing to better energy use in submersible pumps are two factors: the elimination of line-shaft horsepower losses in submersible pumps and a reduction in the amount of horsepower lost due to friction in the column pipe, since submersible pumps do not require an oil tube as shown in *Table 1.* But offsetting those gains are two possible additional factors: horsepower lost in the power cable of submersible pumps and the fact that submersible motors are less efficient than line-shaft motors (see *Table 1*).

Table 1. Comparison of submersible and line-shaft pump performance.

	power transmission losses	***column pipe head losses***	***motor efficiency***
Submersible pump	*0.99 hp/100 ft.*	*1.1 ft./100 ft.*	*89%*
Line-shaft pump	*1.95 hp/100 ft.*	*2.4 ft./100 ft.*	*92%*

Table 2 on the next page compares the horsepower requirements and overall efficiency of submersible and line-shaft pumps. These results show that the overall efficiency of this pumping plant is about the same for both submersible and line-shaft pumps. Thus, factors other than overall efficiency performance may be more important in determining which type of pump to use.

Table 2. Horsepower requirements and overall efficiency of submersible and line-shaft pumps for a pumping lift of 450 feet and a flow rate of 1650 gallons per minute. A pump bowl efficiency of 80 percent is assumed. Bowl setting is 500 feet deep.

	Submersible	*Line-shaft*
Water horsepower[1]	*187.5*	*187.5*
Horsepower losses due to pump inefficiency	*46.9*	*46.9*
Horsepower losses due to column pipe friction	*5.5*	*12.0*
Horsepower losses due to cable or line shaft	*5.0*	*9.8*
Brake Horsepower of motor	*244.9*	*256.2*
Input motor horsepower	*275.2*	*278.5*
Overall efficiency	*68%*	*67%*

[1]*Water horsepower is the power needed to lift water 450 feet at a flow rate of 1650 gallons per minute.*

Reference

Halderman, A.D. 1979. *Submersible vs. line-shaft vertical turbine pumps.* University of Arizona Leaflet 79-1.

Vortexing and Submergence

A vortex can occur when pumping from ponds, reservoirs, etc. A votex is simply a whirlpool and is caused by localized eddies on the water surface that occur when the pump is started. Vortexes can cause the pump to lose prime, reduce head and flow rate, reduce efficiency, and sometimes, cause noise in the pump.

Methods that can help prevent vortexing include the following:

- Increase the submergence of the inlet of the intake pipe. Submergence is the depth of water above the inlet. *Figure 1* shows some recommended values of submergence for various pipe diameters and flow rates. A rule of thumb is that the submergence should be at least 1.5 x D where D is the diameter of the suction bell.
- Use a suction bell at the intake pipe inlet which will reduce inlet losses and help suppress vortexing. For short-coupled vertical turbine pump applications, the use of a suction "umbrella" on the suction bell of the pump will also help to suppress vortexing.
- Float a board on the water surface where vortexing occurs.
- Install baffles and vanes to better guide the water to the pump inlet. This may be appropriate where the pump inlet is in a sump.

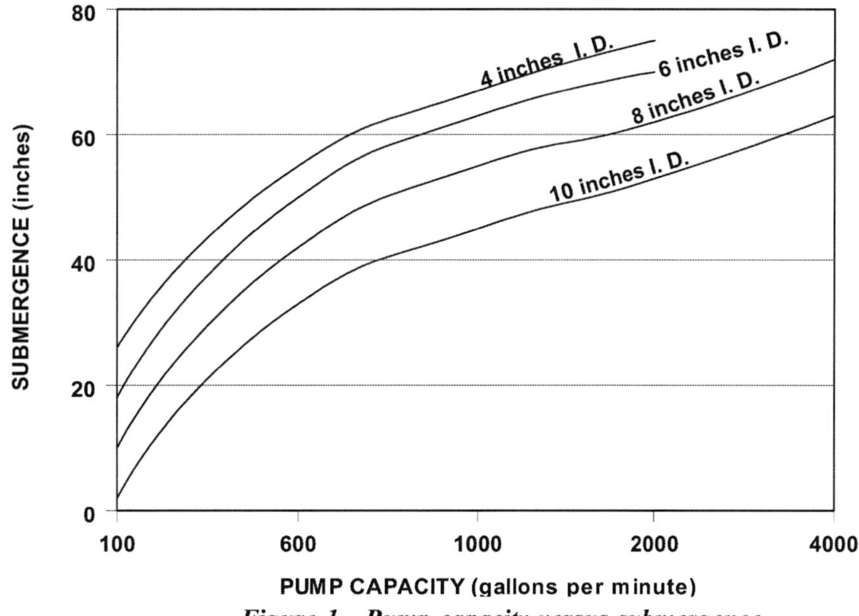

Figure 1. Pump capacity versus submergence.

Department of Land, Air and Water Resources, University of California, Davis

Cavitation

Cavitation is a phenomenon that occurs when the pressure at the impeller intake becomes less than the vapor pressure of water. The vapor pressure of water, which is less than the atmospheric pressure, is that pressure at which liquid water turns to vapor. Under some conditions, the pressure at the entrance to the impeller of centrifugal pumps or impeller inlet may become a vacuum, or less than atmospheric pressure. If the pressure becomes too low, cavitation will occur, and bubbles of water vapor are formed. The bubbles are carried along with the liquid water into the impeller. As the water travels through the impeller, the water pressure starts to increase. At some point, the water pressure causes the bubbles to collapse, which generates a shock wave in the impeller that can cause pitting of the impeller surface. The bubble collapse creates a sound like gravel is inside the pump, and can seriously damage the pump and degrade pump performance.

Cavitation rapidly degrades pump performanc, as shown in *Figure 1*. Once cavitation starts to occur, the total head developed by the pump rapidly drops, thus reducing pump output and efficiency.

Cavitation in centrifugal pumps can be prevented by:

- Avoid excessive suction lift (see "Suction Lift").
- Avoid operating the pump at heads much lower than the head at peak efficiency.
- Avoid operating the pump at capacities much higher than the capacity at peak efficiency.
- Avoid operating at speeds higher than the manufacturer's recommendation.

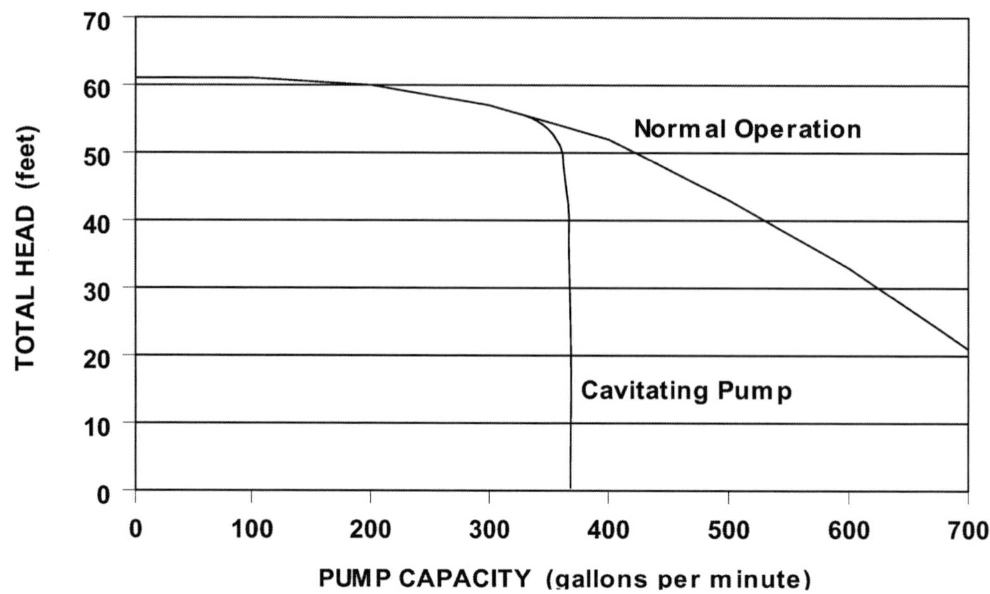

Figure 1. Effect of cavitation on pump performance.

Suction Lift and Net Positive Suction Head

Frequently, centrifugal pumps pumping water out of ditches, ponds, canals, etc. are located above the water surface, thus requiring a suction lift to be developed to lift water from the water surface up to the pump.

What is a suction lift? When a primed (filled with water) pump is turn on, rotation of the impeller reduces the water pressure at the impeller's inlet and causes a vacuum (pressure less than atmospheric pressure) to form. Meanwhile, the water pressure at the surface of the pond, stream, etc. is atmospheric. The difference between the atmospheric pressure and the pressure at the impeller inlet causes water to flow into the intake pipe. This pressure difference must be sufficient to lift the water from the water surface up to the pump and also overcome friction losses in the intake pipe.

A limit exists on the amount of suction lift that can be developed in a pumping plant. If the water pressure at the impeller inlet becomes too low, cavitation will occur. Cavitation occurs when the water pressure equals that of the vapor pressure of water which results in bubbles of water vapor forming in the impeller. As the bubbles of water vapor move through the impeller and are subjected to higher pressures, they collapse very violently and can damage the impeller. When this behavior occurs, pump performance will be severely degraded.

Net positive suction head (NPSH) is a term used to describe the pump's ability to develop a suction lift. There are two types of NPSH. The required NPSH is that which is developed by the pump and depends on pump/impeller design. A performance curve provided by the manufacturer describes the required NPSH for a given pump (*Figure 1*).

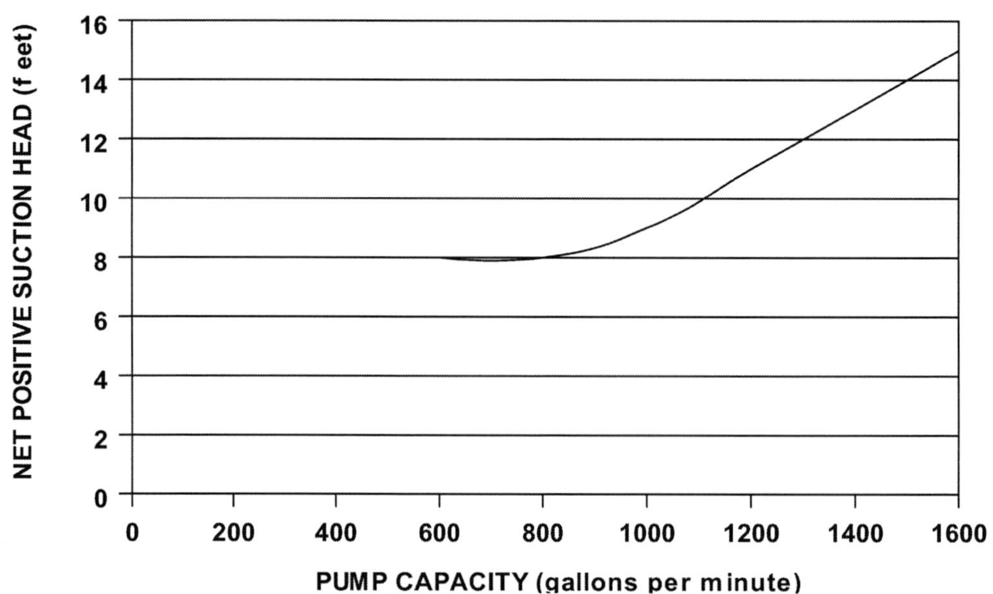

Figure 1. Net positive suction head versus capacity.

The available NPHS depends on the design of the intake of the pumping plant. The available NPSH depends on the elevation difference between pump intake and water surface; friction losses in the intake pipe which depend on pipe diameter, pipe length, flow rate and fittings in the pipe; atmospheric pressure (depend on elevation); and vapor pressure of water (depends on temperature).

Proper operation of the pumping plant means that the available NPSH must exceed the required NPSH. The available NPHS of an existing pumping plant can be determined by first measuring the water pressure at the pump intake and the pump flow rate. A vacuum gauge will be needed to measure the water pressure if the elevation of the pump is higher than the water surface elevation. The available NPSH is then calculated from the following equation:

$$\text{Available NPSH} = H_A + H_V - H_P - H_{VP}$$

where H_A = atmospheric pressure head (feet)
H_V = velocity head of water (feet)
H_P = water pressure head at the pump intake (feet)
H_{VP} = vapor pressure head of water (feet)

The atmospheric pressure head for a given elevation can be determined in *Table 1*. The velocity head can be determined from *Table 2* if the pump flow rate and intake pipe diameter are known. The vapor pressure head is in *Table 3* for various temperatures.

Example

Calculate the available NPSH for the following conditions: Elevation = 500 feet, H_P = 15.1 feet of vacuum, Q = 1000 gpm, D (pipe diameter) = 10 inches, and Temperature = 80°F.

From *Table 1*, the atmospheric pressure head is 33.3 at an elevation of 500 feet; from *Table 2*, the velocity head (H_V) is 0.26 feet; and from *Table 3*, the vapor pressure head is 1.2 feet for a temperature of 80°F .

The available NPSH of this pumping plant is 33.3 – 15.1 + 0.26 – 1.2 = 16.4 feet.

The available NPSH of a proposed pumping plant can also be calculated. However, the water vapor pressure, friction losses in the intake pipe, and elevation differences between pump intake and water surface are required. A discussion of those factors is beyond the scope of this chapter. Those desiring to make this calculation should see a qualified pump dealer.

Table 1. Atmospheric pressure head for various elevations.

Altitude (Feet)	*Atmos. Press. Head (feet)*
-1000	*53.2*
-500	*34.6*
0	*33.9*
+500	*33.3*
+1000	*32.8*
1500	*32.1*
2000	*31.5*
2500	*31.0*
3000	*30.4*
3500	*29.8*
4000	*29.2*
4500	*28.8*
5000	*28.2*
5500	*27.6*
6000	*27.2*
6500	*26.7*
7000	*26.2*
7500	*25.7*
8000	*25.5*
8500	*24.7*
9000	*24.3*
9500	*23.8*
10000	*23.4*
15000	*19.2*

Table 2: Velocity Head (feet) for various pipe diameters and flow rates.

Pump Flow Rate (gallons per minute)	*Pipe Diameter (inches)*					
	6	*8*	*10*	*12*	*14*	*16*
100	*0.02*	*0.01*	*0.00*	*0.0013*	*0.0007*	*0.0004*
200	*0.08*	*0.03*	*0.01*	*0.01*	*0.003*	*0.002*
300	*0.18*	*0.06*	*0.02*	*0.01*	*0.01*	*0.00*
400	*0.32*	*0.10*	*0.04*	*0.02*	*0.01*	*0.01*
500	*0.50*	*0.16*	*0.07*	*0.03*	*0.02*	*0.01*
600	*0.73*	*0.23*	*0.09*	*0.05*	*0.02*	*0.01*
800	*1.29*	*0.41*	*0.17*	*0.08*	*0.04*	*0.03*
1000	*2.01*	*0.64*	*0.26*	*0.13*	*0.07*	*0.04*
1200		*0.92*	*0.38*	*0.18*	*0.10*	*0.06*
1400			*0.51*	*0.25*	*0.13*	*0.08*
1600			*0.67*	*0.32*	*0.17*	*0.10*
1800			*0.85*	*0.41*	*0.22*	*0.13*
2000				*0.50*	*0.27*	*0.16*
2200				*0.61*	*0.33*	*0.19*
2400				*0.73*	*0.39*	*0.23*
2600				*0.85*	*0.46*	*0.27*
2800				*0.99*	*0.53*	*0.31*
3000				*1.13*	*0.61*	*0.36*

Table 3. Vapor pressure head of water.

Temperature (°F)	*Vapor Pressure Head (feet)*
60	*0.59*
70	*0.89*
80	*1.2*
85	*1.4*
90	*1.6*
100	*2.2*
110	*3.0*
120	*3.9*
130	*5.0*

Reference

Garay, Paul N. 1993. Pump application desk book. The Fairmont Press. 471 p.

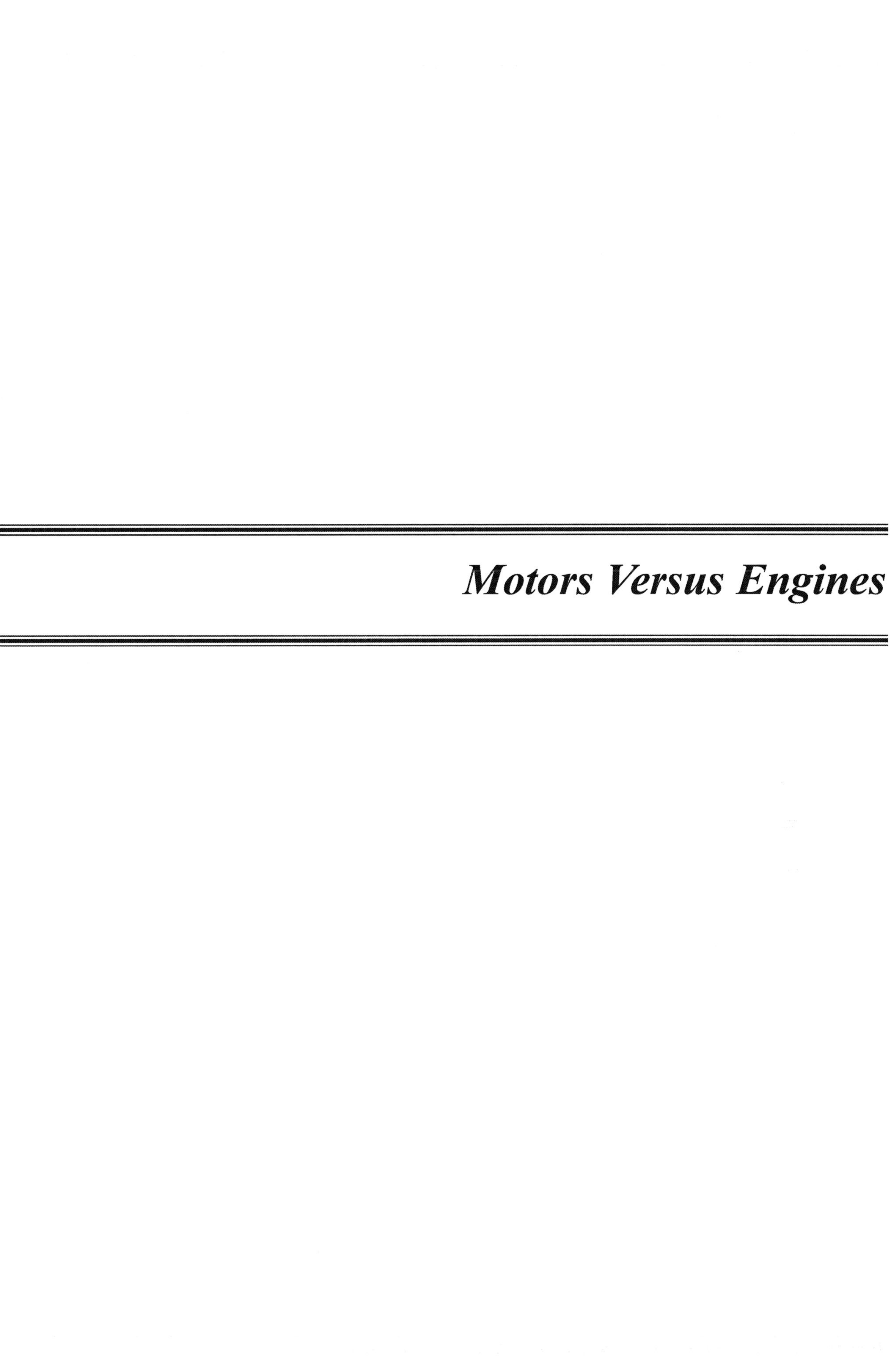

Motors Versus Engines

Electric Motors

Many pumping plants in California use electric motors, which require minimal maintenance and are very reliable. The specific characteristics of electric motors are given below.

• Electric motors operate at relatively constant efficiency and rpm under a wide range of loads. *Table 1*, which gives motor efficiencies for various horsepowers, shows a slight decrease in motor efficiency when the motor load is decreased from 100 percent to 50 percent of the rate horsepower. For example, the efficiency of a 15-horsepower motor is 89 percent at 100 percent-rated load and 85 percent at 50 percent load, but as motor horsepower increases, the effect of the load on efficiency decreases. The motor efficiency decreases substantially for motor loads less than 50 percent of the rated horsepower.

Table 1. Motor efficiency (%) at various horsepower and loads (% of rate horsepower).

Horsepower	*Efficiency*		*Horsepower*	*Efficiency*	
	100% load	*50% load*		*100% load*	*50% load*
3	84	81	60	90	87
5	85	80	75	90	90
7.5	86	81	100	90	89
10	87	83	125	91	90
15	89	85	150	91	90
20	89	85	200	92	90
30	89	83	250	92	91
40	90	89	300	92	91
50	90	89	350	92	91
			400	92	91

• Larger motors are more efficient than smaller motors.
• Motor efficiency does not change substantially with age.
• Motors draw power in proportion to the load. Applying a load more than the rated horsepower will cause the motor to draw more horsepower than its rated power.

• The maximum load a motor can continuously tolerate is expressed by the service factor, which is listed on the nameplate. Most service factors are 1.15, meaning that the motor can tolerate a 115 percent load. The actual load can be calculated by the following formula:

$$\text{load} = (\text{IHP} \times \text{Em}) / (100 \times \text{nameplate horsepower})$$

where: IHP = input horsepower (measured from a pump test)
Em = motor efficiency (%)

Energy Efficient Motors

Energy efficient motors are designed to operate at a higher efficiency compared with standard motors, as shown in *Table 2.*

Table 2. Efficiencies (%) of a standard motor and an energy-efficient motor.

HP	*Standard*	*Energy Efficient*
10	86.5	91.7
20	86.5	93.0
50	90.2	94.5
75	90.2	95.0
100	91.7	95.8
125	91.7	96.2

The higher efficiencies are achieved by using improved material and design. The service factor of these motors is the same as that of standard motors. Energy efficient motors also maintain a higher efficiency when underloaded compared to standard motors. Energy efficient motor cost more than do standard motors. The cost differences for one motor manufactures ranged from 20% to 24% more for the energy efficient motor. Thus, does converting to energy efficient motors pay?

One method to answer this question is to calculate the payback period. This is determined by dividing the cost difference between an energy-efficient motor and a standard motor by the savings achieved by converting to a more efficient motor. The savings is calculated by:

$$S = 0.746 \times HP \times C \times T \left(\frac{100}{SE} - \frac{100}{EE} \right)$$

C = energy costs ($/Kwhr)
T = operating hours
SE = standard motor efficiency
EE = energy efficient motor efficiency
HP = rated HP of motor

A payback period less than 2 years is excellent, a 2-4 payback period is good while a period of more than 5 years is poor.

In general, an energy-efficient motor should be used for new pumping plants or when the standard motor needs to be replaced due to wear or breakdown. However, simply exchanging a standard motor with an energy efficienct one will not likely be economical.

Example: What is the payback period for purchasing an energy-efficient motor to be used in a new irrigation pumping plant.

C = \$0.11 kwhr
T = 3000 hr
Horsepower = 100
SE = 91.7%
EE = 95.8%
Cost of standard motor = \$4,900
Cost of energy efficient motor = \$6,050

$$S = 0.746 \times 100 \text{ HP} \times \$0.11 \text{ kwhr} \times 3000 \text{ hr} \left(\frac{100}{91.7\%} - \frac{100}{95.8\%} \right)$$

= \$1,149 per year

Cost of difference between motors = \$6,050 - \$4,900 = \$1,150

Payback period = \$1,150 / \$1,149 = 1 year

This payback period is considered to be excellent.

Example: What is the payback period for simply exchanging an existing standard motor with an energy efficient motor using the data of the previous example.

Savings = \$1,149 per year

Cost of energy-efficient motor = \$6,050

Payback period = \$6,050 / \$1,149 = 5.3 years

Some utility companies may subsidize the extra cost of an energy efficient motor. For this case, the payback period is:

Payback period = \$4,900 / \$!,149 = 4.3 years

Factors Affecting Motor Performance

Too much heat can greatly reduce motor life by damaging the insulation of the motor coil windings. The following factors can cause excessive heat:

• Motor overload;

• Poor ventilation, preventing adequate air circulation throughout the motors;

• Low voltage, causing the motor to draw more amperes to maintain both motor speed and horsepower;

• Voltage imbalance among the three phases of a three-phase motor.

Ways to Increase Motor Life

The following measures can increase motor life:

• Keep ventilation openings clear;

• Maintain motor load at 100 percent or less of rated horsepower;

• Keep voltage imbalance between phases at 4 percent or less;

• Increase power factor (as measured by a qualified electrician). The power factor is the fraction of total current effective in producing power.

References

Hansen, H.J. and W.L. Trimmer. 1986. *Extending electric motor life.* Pacific Northwest Extension Publication No. 292.

Knutsen, G., R. Curley, and W. Chancellor. 1981. *Selecting electric motors for maximum efficiency.* University of California Leaflet 21240.

Schroeder, M.A. 1982. "Three-phase electrical motors." In *Irrigation Pumping Plants and Well Efficiency Handbook.* University of Nebraska, Lincoln, Nebraska.

Variable Frequency Drives for Electric Motors

Electric motors used for irrigation pumping plants usually operate at a constant rpm. The pump chosen for a pumping plant should provide the design capacity and total head at maximum efficiency for a given rpm. But actual operating conditions may vary considerably from the design conditions, as when a single pump is used to irrigate fields of different sizes. Under those circumstances, the pump discharge can be throttled by partially closing a valve in the discharge pipe to lower the pressure and reduce the pump output to the smaller fields. Throttling may slightly decrease the horsepower demand of the pump, but will cause substantial amounts of energy to be lost across the valve, resulting in low efficiency.

One solution to these problems is to use a variable or adjustable frequency drive for electric motors. Variable frequency drives allow the rpm of the motor to be reduced by varying the frequency of the power into the motor, which in turn reduces the motor horsepower. The drives consist of a converter that changes AC power to DC power and an inverter that changes the DC power into adjustable frequency AC power. As the frequency of the power is decreased, the power to the motor and the motor rpm are both reduced. This decrease in motor rpm can substantially reduce the pump horsepower demand, since the pump horsepower demand is proportional to the pump rpm^3—with the result that a small change in rpm causes a significant change in pump horsepower demand.

Figure 1 on the next page illustrates the relationship between rpm reductions and horsepower reductions. This relationship was developed from equations known as *affinity laws,* which describe how changes in rpm affect total head, capacity, horsepower, and efficiency. The figure shows that reducing the rpm by about 20 percent will reduce the horsepower demand by about 50 percent. Reducing the rpm from 1770 down to 1400, for example, will decrease the horsepower demand of a 100-horsepower pump to 50 horsepower.

Matching Pump Output to Rpm

The pump output — which consists of the pump capacity and the total head — is also determined by the rpm. The capacity is proportional to the rpm, while the total head is proportional to the rpm^2. *Figure 1* also illustrates these relationships. A 20% change in rpm will decrease the pump capacity by 20% and the total head by nearly 38%.

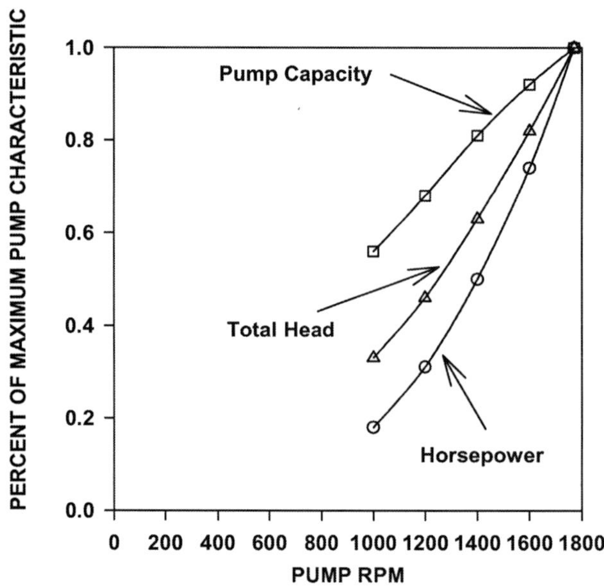

Figure 1. Ratio of pump characteristics to pump rpm.

Because of these relationships, adjusting the pump rpm may not yield the same total head and capacity obtained under the throttled conditions. The actual total head and capacity at a particular rpm will depend on the impeller design, which defines the relationship between total head and pump capacity.

Effect on Motor Temperature

Some concerns exist about the effect of variable speed drives on motor temperature, since the adjustable AC power generated from DC power may contain distortions not found in AC power. An increase in temperature can shorten the life of an electric motor. A report published by the Electric Power Research Institute, *Adjustable Speed Drives: Application Guide,* confirms that a variable speed drive can raise motor temperature by 3 to 15 percent, but adds that variable speed drives of more recent design produce less than a 5 percent temperature increase.

Effect on Pumping Plant Efficiency

A variable frequency drive can also affect the efficiency of the pumping plant. The lower the rpm, the less efficient the motor and the variable frequency drive. Down to about 50% of the maximum rpm, the drive efficiency may decrease only slightly, but at lower rpms the efficiency of the drive falls dramatically. The EPRI report cited above estimates drive efficiency at 50% speed to be about 84%, compared to 95% at 100% rpm, while at 25% of maximum rpm, the drive efficiency falls to 53%. Pump efficiency, however, is affected only slightly by rpm changes.

Example

A centrifugal pump was used to irrigate blocks of tree crops of different acreages with water supplied from a reservoir. The pump was tested first under constant rpm while irrigating an 80-acre block and next while irrigating a 50-acre block, which required throttling the pump discharge. A third test was conducted using the variable speed drive to reproduce the output of the throttled pump.

Table 1 shows the results of the pump tests. The variable speed operation at 1345 rpm adequately reproduced the output of the throttled operation (rpm = 1770). There was little pressure difference between the two operating conditions, but the variable speed operation produced 100 gpm more. The lower pumping plant efficiency of the throttled operation reflects energy lost because of the partially closed valve. The input horsepower demand of the electric motor decreased from 90 horsepower under the throttled condition down to 55 horsepower under the variable speed condition. Note that the pumping plant efficiency of the variable speed condition is nearly the same as that of the unthrottled operation, indicating that pumping plant efficiency is not affected by varying the rpm.

Table 1. Pump test results—tree crops.

	Unthrottled	*Throttled*	*Variable Speed*
Acres	*80*	*50*	*50*
Pressure (psi)	*80*	*64*	*60*
Pump capacity (gpm)	*1100*	*600*	*700*
Input horsepower	*128*	*90*	*55*
RPM	*1770*	*1770*	*1345*
Overall efficiency (%)	*40*	*22*	*44*

Summary

Variable frequency drives offer the potential of adjusting the output of an irrigation pump to match changing operating needs while also reducing energy costs. Several field trials including the one reported herein have shown that pump output can be adequately matched to desired output by lowering the frequency of the power into the motor, thereby reducing the pump rpm. In these tests, the reduction in horsepower ranged from 25 to 64 and in most cases was accomplished by reducing the pump rpm from 1770 down to about 1400.

The economic analysis of these trials suggests that the pump should be operated at the reduced horsepower for at least 1000 hours per year to provide annual positive net returns.

Note

The author is indebted to the following individuals who provided information for this chapter: Matt Kehoe, Central Pump, Modesto, California; Cris Hepburn, McCalla Water Well Services, Redlands, California; and Cathy Calloway, Campbell and George Company, San Carlos, California, as well as to the Electric Power Research Institute, Pleasant Hill, California for providing the reports, *Adjustable Speed Guides: Applications Guides* and *Adjustable Speed Drives: Directory*.

Engines

When an engine is to be used for a pumping plant, it must match the needs of the pump. Choosing the right engine for a pumping plant is more difficult than choosing an electric motor because the performance of an engine — brake horsepower, fuel consumption, and overall efficiency — are all affected by the engine rpm.

Brake Horsepower and Rpm

Figure 1 shows that the higher the rpm the greater the engine brake horsepower (the shaft horsepower of the engine). At 1200 rpm, the brake horsepower is 110; at 2000 rpm, the brake horsepower is 167.

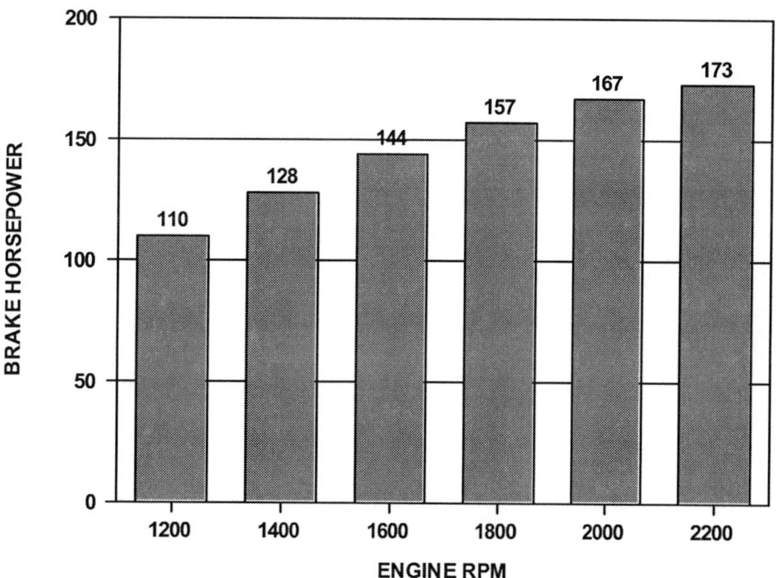

Figure 1. The relationship between engine brake horsepower and rpm.

Fuel Consumption, Engine Efficiency, and Rpm

Unit fuel consumption also changes with rpm. *Figure 2* on the following page shows that fuel consumption decreases to a minimum at 1600 to 1800 rpm and increases at higher rpms.

Figure 3 shows that engine efficiency is also affected by rpm, increasing to a maximum of 35.1 percent at 1600 to 1800 rpm and decreasing at higher rpms.

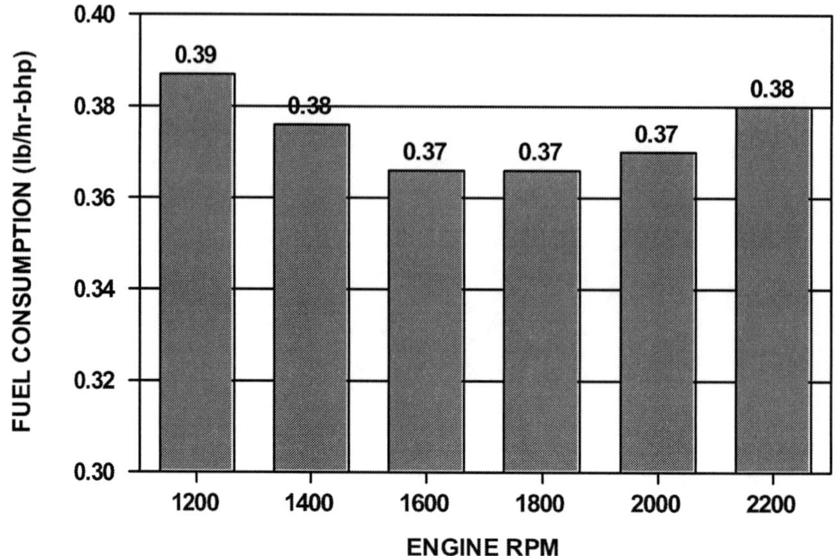

Figure 2. The relationship between unit fuel consumption and rpm.

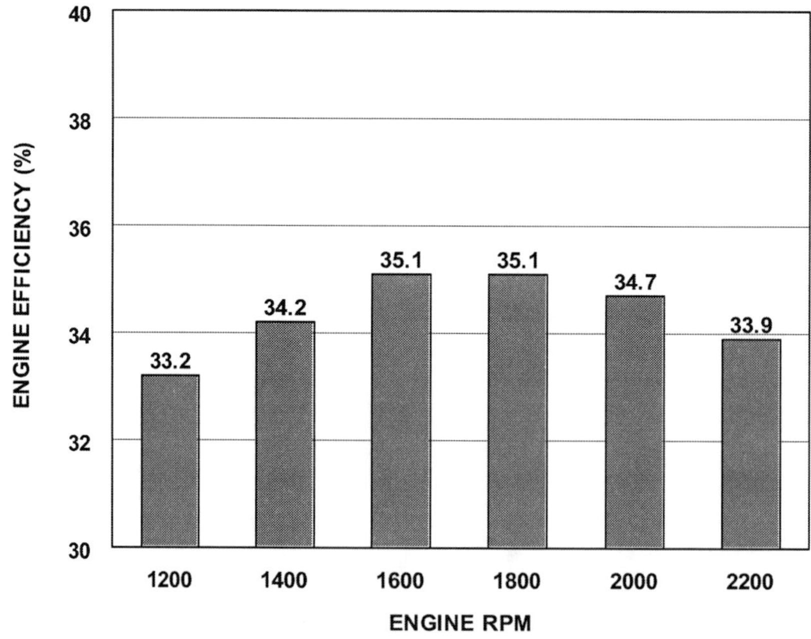

Figure 3. The relationship between engine efficiency and rpm.

For a pumping plant to perform efficiently, therefore, the engine must provide the required brake horsepower — that needed by the pump — at its point of maximum efficiency. The rpm at maximum efficiency must also be appropriate for the pump rpm. Pump head gear ratios may have to be adjusted to match rpm requirements. A dealer can help accomplish this purpose.

The choice of which engine to use in a pumping plant must be based on the pumps's continuous, *derated* horsepower, not on its maximum horsepower. Engine horsepower is affected by a number of factors, including the equipment installed on the engine, climatic conditions, elevation, and engine wear. The following guidelines may be useful in matching an engine to a pump:

Guidelines for Matching an Engine to a Pump

- The continuous operating horsepower of an engine is 70%-75% of its maximum horsepower.
- The engine horsepower available for pumping must be estimated by derating the engine — adjusting the horsepower to allow for climatic conditions and additional engine components. When engines of different manufacturers are compared, each engine should have identical equipment installed during testing. Further derating may be required to allow for fans, belt drives, alternators, and exhaust systems if this equipment was not installed on the engine during manufacturer tests. Additional derating may be required to determine temperature and elevation effects. Cooperative Extension agricultural engineers at UC Davis can assist in comparing engines.
- Engine horsepower decreases about 1.0 percent for every 10 degrees above 85 degrees.
- Engine horsepower decreases about 1.5 percent for every 500 feet above sea level. Turbocharged engines may decrease by about 1.0 percent for every 500 feet above sea level.
- Engine wear will reduce horsepower. Good maintenance will slow wear.

Converting From Motors to Engines

Advantages

Whether a pumping plant should be converted from an electric motor to an engine depends on relative cost, which in turn is influenced by factors specific to the pumping plant site. The first step in making the decision is to compare energy costs for electricity and fuel. The following equations can help provide a fairly accurate comparison:

Electric: $/AF = 1.58 x Total head x $/Kwhr
Engine: $/AF = 0.139 x Total head x $/gallon

where: $/AF = dollars per acre foot of water pumped,

Total head = total lift developed by pumping plant in feet (obtained from pumping plant test),

$/Kwhr = electrical energy costs (includes demand charges),

$/gallon = engine fuel costs

These equations assume a motor efficiency of 90 percent, an engine efficiency of 25 percent, and a pump efficiency of 72 percent (this differs from pumping plant efficiency, which includes the motor or engine efficiency.)

Compare energy costs using the following information:

Total head or lift = 100 feet
$/Kwhr = 0.10
$/gallon = 0.70

Electric motor: $/AF = 1.58 x 100 feet x $0.10 Kwhr
= $15.8/AF

Engine: $/AF = 0.139 x 100 feet x $0.70/gallon
= $9.7 AF

According to this analysis, the energy costs of using engines would be less than the energy costs of using electric motors.

The next step is to compare the total annual capital, maintenance, and energy costs. Comparing these costs can be complicated, but a computer program designed for this purpose is available from the Department of Biological and Agricultural Engineering, University of California, Davis, CA 95616.

Matching Engine to Pump Horsepower

Both engine horsepower and pump horsepower are affected by rpm. The higher the pump rpm, the more horsepower demanded by the pump, and the higher the engine rpm, the more horsepower developed by the engine. The engine chosen should be one that develops the needed horsepower at maximum engine efficiency, but the rpm at which the needed horsepower is developed may be different from the pump rpm.

Differences in rpm can be overcome by choosing a pump gear head with the appropriate gear ratio. *Figure 1* shows the effect of rpm on pump and engine horsepower. At about 1770 rpm, the pump requires 72 horsepower, but the engine develops 72 horsepower at about 2000 rpm. A gear ratio of 5:6 could help match the engine and pump (1770 ÷ 2000 = 5/6).

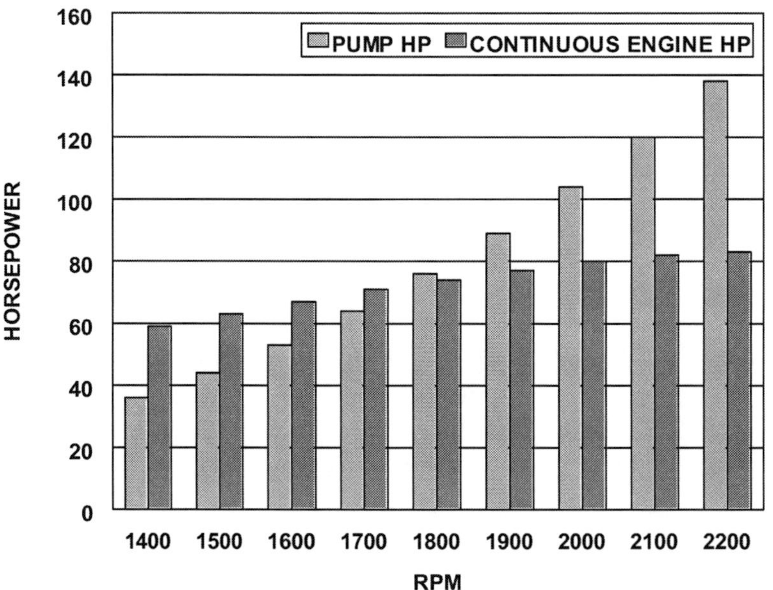

Figure 1. The relationship between pump horsepower, engine horsepower, and rpm.

Conditions to Avoid

Because engine rpm can be varied, the temptation might be to increase pump rpm to increase pump flowrate and total head, particularly if groundwater levels have been declining. But as rpm increases, pump horsepower increases at a much faster rate than engine horsepower, so that increasing the rpm can overload the engine. For example, if the rpm is increased to 2200 rpm, the engine depicted in *Figure 1* will develop 76 horsepower, while the pump will require 138 horsepower. This will severely overload the engine and greatly decrease its life.

Engine rpm should not be increased without considering the relationship between rpm and the pump horsepower demand.

Reference

Knutsen, Jerry. 1981. "Converting from electrical power to diesel power for irrigation." Department of Agricultural Engineering, University of California, Davis.

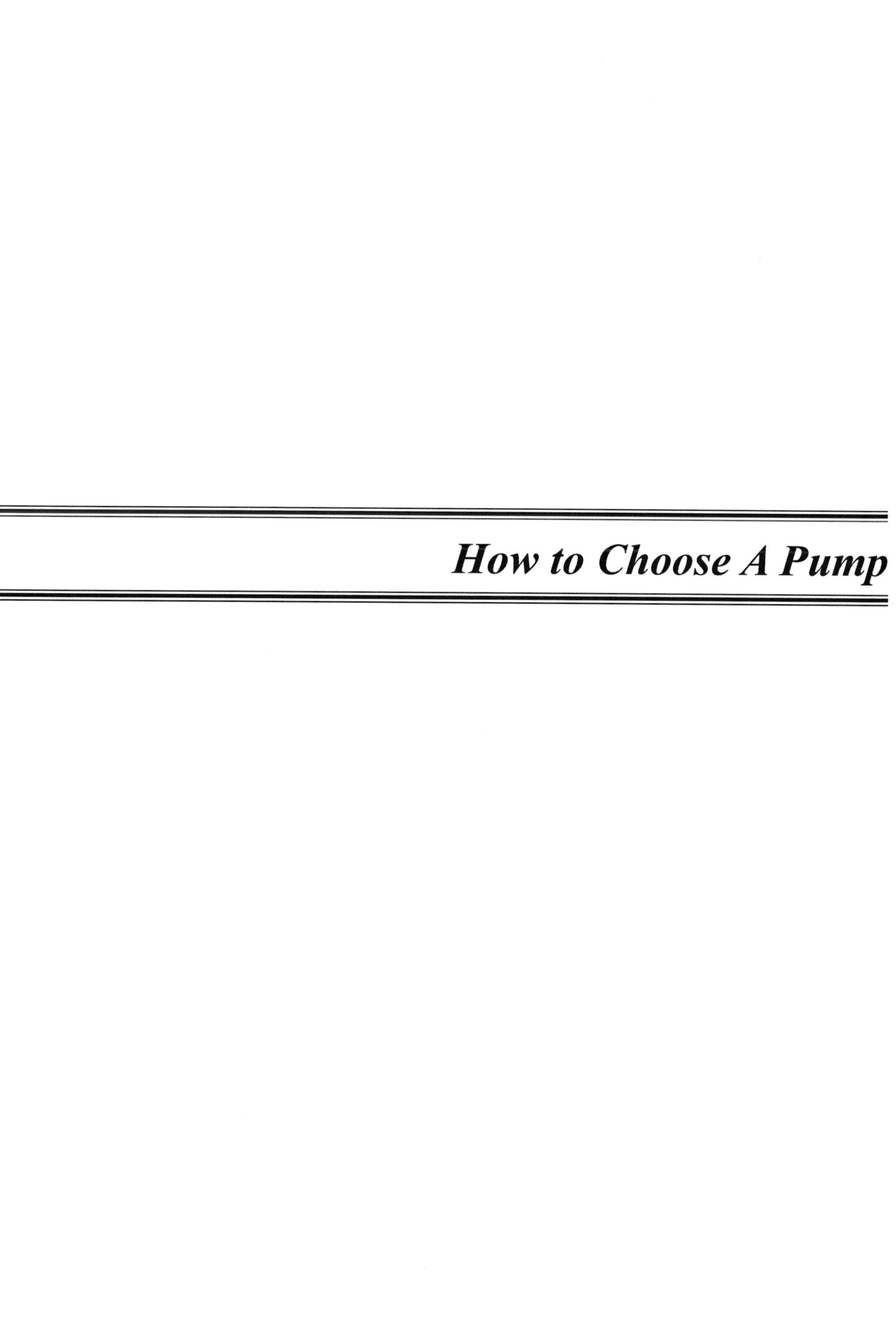

How to Choose A Pump

What to Consider in Choosing a Pump

The task in choosing a pump is to find a pump that provides the desired capacity and total head while operating at close to maximum efficiency — that is, while using the least possible horsepower and energy and incurring the lowest possible demand charges.

Capacity

The capacity of the pump must satisfy the crop water needs during the period of maximum evapotranspiration (crop water use). The capacity needed depends on the crop evapotranspiration rate, the number of acres being irrigated, the efficiency of the irrigation (how much water is used by the crop, compared to how much evaporates, is lost to runoff, or percolates down past the root zone), how many hours the pump operates at each irrigation, and how long the interval between irrigations. The capacity can be estimated by the following formula:

$$Q = \frac{(449)\ (A)\ (d)}{t}$$

where :

Q = capacity (gpm)
A = acres irrigated
d = desired depth to apply during an irrigation (the crop evapotranspiration divided by the irrigation efficiency) in inches
t = pump operating time needed for an irrigation (hours).

Example: Calculate the capacity needed to apply four inches over 100 acres. 200 hours are required for an irrigation.

$$Q = \frac{(449)\ (100\ \text{acres})/(4\ \text{inches})}{200\ \text{hours}} = 898\ \text{gpm}$$

Total Head

Total head consists of the pumping lift and the discharge pressure head. The pumping lift is the distance in feet required to raise the water to the ground surface. The pump discharge pressure is the pressure needed by the sprinklers or emitters in addition to the pressure needed to overcome elevation differences and pressure losses caused by friction. The total head is the sum of the pumping lift and discharge pressure head (pressure in psi multiplied by 2.31).

Selection Procedure

Following, then, are the steps to take in choosing a pump:

- Estimate the needed capacity and total head.
- Use a catalog of manufacturer's pump performance curves to find a pump that will provide the needed capacity and total head at close to maximum efficiency.
- Use the pump performance curve to determine the horsepower requirement of the pump.
- Select the appropriate electric motor or engine. The engine rpm should be close to maximum engine efficiency.

Factors to Consider in Making the Selection

The following factors should be taken into account:

- ***Changing groundwater levels.*** Changes in groundwater level can affect pump capacity. Select a pump with a relatively steep total head/capacity curve to cut down changes in pump capacity caused by fluctuating water levels.
- ***Varying sprinkler laterals.*** Changing the number or length of sprinkler laterals during irrigation will alter both pressure and capacity. Selecting a flat total head/capacity curve will cut down on pressure variations.
- ***Choosing the right size for the deep-well turbine.*** Pump size is limited by the diameter of the well casing. "Sizing the Deep-Well Turbine", page 65, lists pump sizes recommended for various casing diameters.

Coping With Changing Irrigation Needs

Often irrigators are faced with a variety of irrigation needs, as when fields are of different sizes, where soil type varies, where more than one kind of crop is grown, or where different types of irrigation systems are used throughout the farm. Any of the following strategies might be followed in choosing a pumping plant to meet these varying circumstances.

- Choose a single pump capable of meeting all the various pumping needs. If this strategy is followed, the pump output (particularly the capacity) should match the maximum irrigation requirement. A valve may be required in the pump discharge pipe to lower both pressure and capacity when less output is needed, as for smaller fields. Throttling the pump output with a valve will not overload an electric motor, but will be inefficient. It may be desirable to choose a pump with a flat total head/ capacity curve to allow for a wide range of capacities at a relatively constant pressure.

- Include several centrifugal or booster pumps in a variety of sizes depending on the range of irrigation needs. Where the water source is groundwater, using only one deep-well turbine or submersible pump is usually practical. The pump should be of a size to provide pumping lift at at least maximum capacity. A deep-well turbine may have to be throttled to match the performance of the booster pumps used when less output is needed.

- Use a variable-frequency drive or engine to change the pump output as needed. (*See* "Variable Frequency Drives for Electric Motors", page 47, for a detailed discussion discussion about these drives.)

Choosing a Pump: Example

Select a pump that will provide 940 gpm, a discharge pressure of 50 psi, and a pumping lift of 112 feet. The pump must provide a total head of 228 feet (50 psi x 2.31 + 112 feet). Use the performance curves for the three pumps shown in *Figure 1*, below.

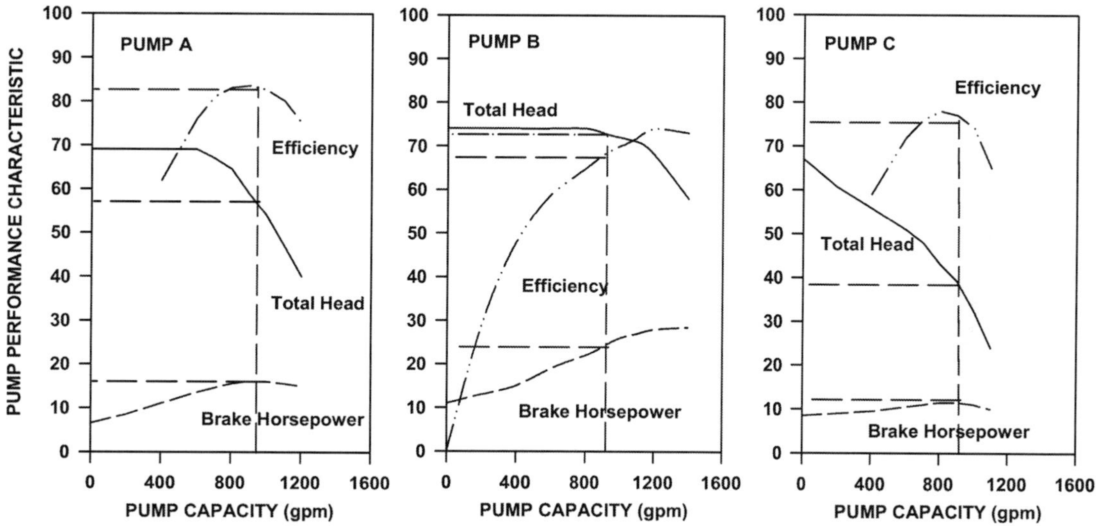

Figure 1. Examples of performance curves for three pumps.

Table 1 shows the total head, brake horsepower, pump efficiency at 940 gallons per minute, and the number of stages for each pump. The total energy use for 1500 hours of operation at $0.10 per kwhr is also shown. Pump B requires three stages, while Pump C needs six stages. Pump A is the most efficient and Pump B is the least efficient.

Table 1. Total head, brake horsepower, pump efficiency, number of stages, and annual energy cost at 940 gallons per minute for three sample pumps.

	A	*B*	*C*
Number of stages	*4*	*3*	*6*
Total head (feet)	*228*	*219*	*223*
Capacity (gpm)	*940*	*940*	*940*
Brake horsepower	*64*	*73.5*	*69*
Pump efficiency (%)	*84*	*69*	*76*
Annual energy cost ($)	*7162*	*8225*	*7721*

Pump A is the best pump for the given conditions. Four stages will develop 228 feet of total head at the highest pump efficiency, requiring the least horsepower and resulting in the lowest annual energy cost.

Note that while Pump B will develop the required total head, the total head/capacity performance curve is flat near this operating point, meaning that slight increases in total head, such as those caused by declining groundwater levels, will cause significant decreases in pump capacity. In contrast, Pump C has a steep total head/capacity curve, meaning that changes in total head will cause less change in pump capacity.

Sizing The Deep-Well Turbine

Pump size is limited by the diameter of the well casing. If the pump is too large for the casing, rocks from the gravel pack can become lodged between the casing and — particularly if the well is crooked — the pump may be difficult to install.

Table 1 provides recommended pump sizes for various well casing diameters.

Table 1. Recommended pump sizes for various well casing diameters.
(*Source:* Driscoll, F.G. 1986. *Groundwater and Wells.* Johnson Filtration Systems, p. 415.)

Anticipated Well Yield (gpm)	*Optimum Size of Well Casing (inches)*	*Smallest Size of Well Casing (inches)*	*Nominal Size of Pump Bowls (inches)*
Less than 100	6 ID	5 ID	4
75 to 175	8 ID	6 ID	5
150 to 350	10 ID	8 ID	6
300 to 700	12 ID	10 ID	8
500 to 1000	14 OD	12 ID	10
800 to 1800	16 OD	14 OD	12
1200 to 3000	20 OD	16 OD	14

Note: ID = inside diameter; OD = outside diameter. The size recommended for the well casing is based on the outer diameter of the bowls for deep-well turbines (line shaft) and on the diameter of either pump bowls or motor for submersibles.

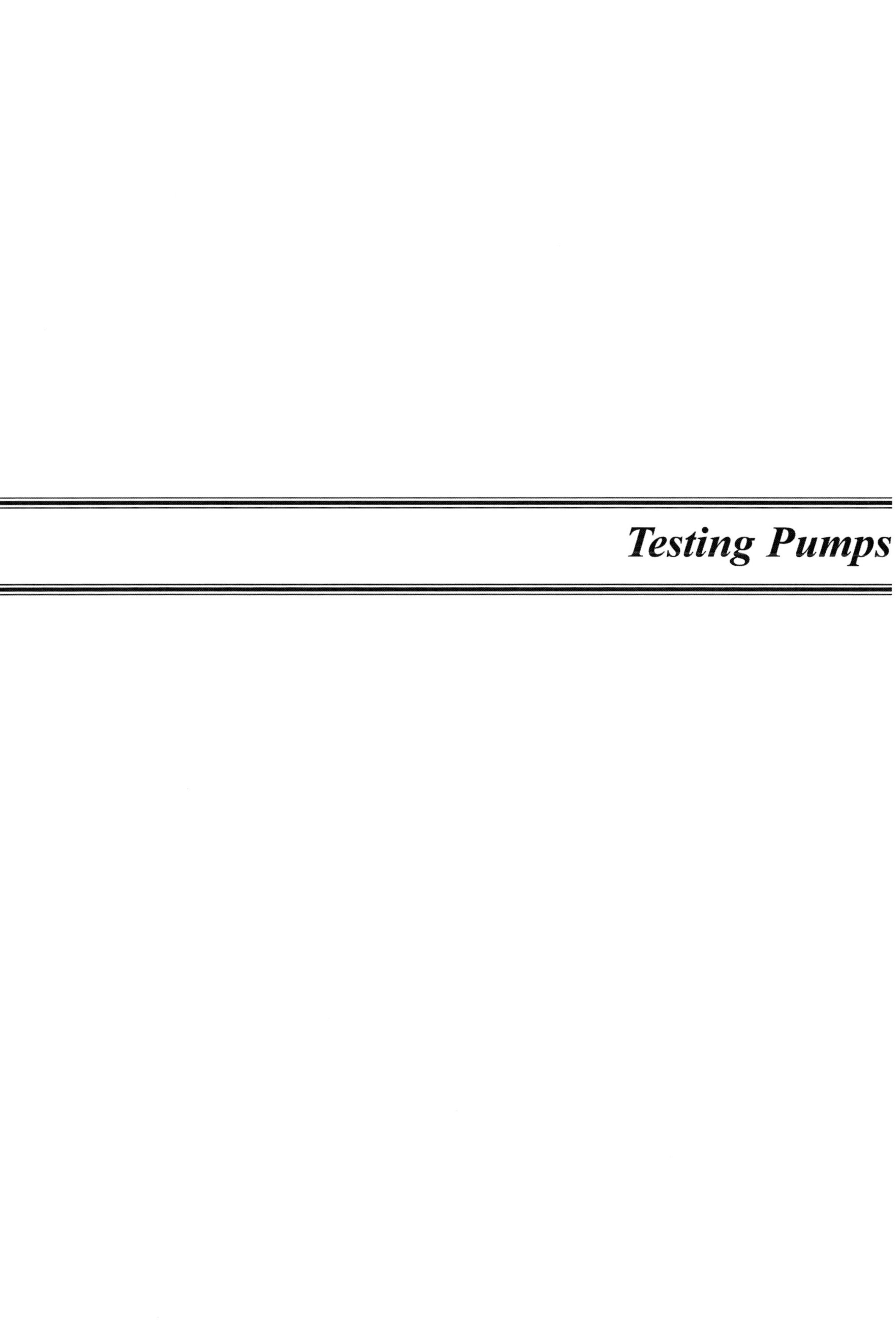

Testing Pumps

Pump Performance Tests

Deep-well turbine and centrifugal pumps — as pump performance curves illustrate — will operate under a range of conditions from maximum capacity to shutoff. An initially efficient pump with the necessary head and capacity can become inefficient because of changed pumping conditions, wear, or other factors. When that happens, the pump can be tested in a two-step process that includes gathering information about the pump's current performance and then comparing that information to the manufacturer's performance curves.

The pump test consists of measuring total head, capacity, and input power to the pump and then calculating overall pumping plant efficiency — that is, the combined efficiency of the pump and motor or engine. The overall efficiency will be less than the pump efficiency reflected in the manufacturer's performance curve because of the efficiency of the motor or engine.

What a Pump Test Covers

Pump tests are conducted by utility companies, pump dealers, and consultants. In defining the current status of the pump, the test may provide any or all of the following information:

- pumping lift
- static or standing water level
- pumping water level
- drawdown
- discharge pressure
- discharge pressure head
- pump capacity
- well yield (a well yield decreasing over time may indicate plugging in the well)
- acre-feet pumped per 24 hours
- output horsepower to motor
- % motor overload
- input kilowatts to motor
- kilowatt-hours per acre-foot of pumped water
- overall pumping plant efficiency

The following page explains more fully some of the measurements covered by a pump test.

• ***Pumping lift.*** The pumping lift is the height between the discharge pipe at the pump head and the water level in the pumping well. Pumping lift is measured by lowering an electric water level indicator (two electrodes attached to a wire connected to an ampere meter) into the space between the well casing and the column pipe. In some cases, an air line is permanently installed to measure the pumping lift. This measurement requires access to the inside of the well casing; without such access, the pump test cannot be performed.

• ***Discharge pressure.*** The discharge pressure is measured with a pressure gauge. The pressure is converted to discharge pressure head by the following equation:

$$\text{discharge pressure head (feet)} = 2.31 \times \text{pressure (psi)}$$

• ***Total head.*** The total head is the pumping lift added to the discharge pressure head.

•***Pump capacity.*** The pump capacity is measured with a flowmeter installed in the discharge pipe. To operate properly, a flowmeter requires a straight section of pipe eight to ten pipe diameters long immediately upstream from the meter to prevent or reduce turbulence in the water (*Figure 1*). Too much turbulence from valves, bends, or elbows will prevent accurate flow measurements. It is also recommended that a straight section of pipe two pipe diameters long be installed immediately downstream from the flowmeter.

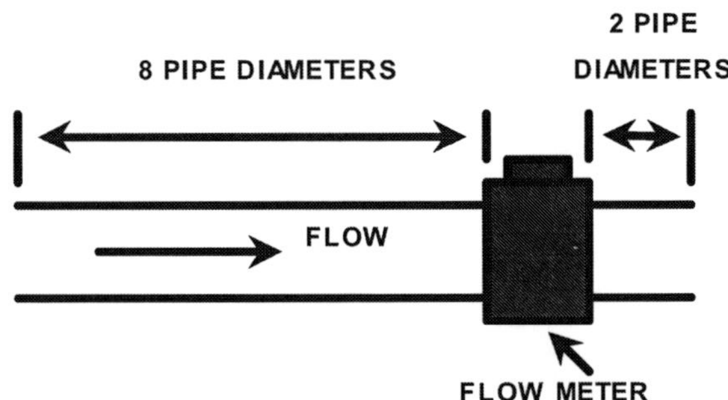

Figure 1. Recommended flow meter installation configuration.

• ***Input horsepower.*** The input power of an electric motor can be determined from the pump power meter by counting the time for the disc inside the meter to complete ten revolutions and by the following equation:

$$\text{IHP} = \frac{(4.81)(10)(\text{Kh})}{t}$$

where:

IHP = input horsepower
Kh = meter constant, found on the faceplate of the meter
t = time in seconds for ten revolutions

• ***Overall pumping plant efficiency***. Overall efficiency can be calculated by the following equation:

$$E = \frac{(Q)(H)}{(3960)(IHP)}$$

where:

E = overall efficiency
Q = capacity (gpm)
H = total head (feet)

Once the pump test has defined the pump's current status, the pump test data can be compared to the manufacturer's performance curves to determine how the pump's actual performance compares to the manufacturer's recommended performance. This comparison step is important because it is possible to have a pumping plant measuring low efficiency but nonetheless operating properly. Repairing such a pump would be of no benefit.

The comparison procedure is to plot the test data on the graph used for the performance curve. First, calculate the head per stage by dividing the measured total head by the number of stages and draw a horizontal line across the graph at the value equal to the measured total head per stage. Next, find the measured capacity along the bottom of the curve and draw a vertical line corresponding to that value. The intersection of the two lines is the operating point of the pumping plant. If the operating point is close to the head/capacity performance curve, the pump is operating properly, regardless of the overall efficiency calculated from the test data. If there is a significant difference between the performance curve and the operating point, the pump may be worn and should be repaired or replaced. (Note that if the pump is situated deep in the well, pressure losses caused by friction in the column pipe may have to be added to the total head.)

Example

Figure 2 on the next page shows the manufacturer's curve for total head, capacity and different operating points obtained from two tests conducted on a particular pump — the first when the pump was worn, with an overall efficiency of 39 percent and the second after the pump was repaired, with an overall efficiency of 50 percent. The worn pump showed a significant difference between the manufacturer's curve and the operating point, while the repaired pump showed only a slight difference.

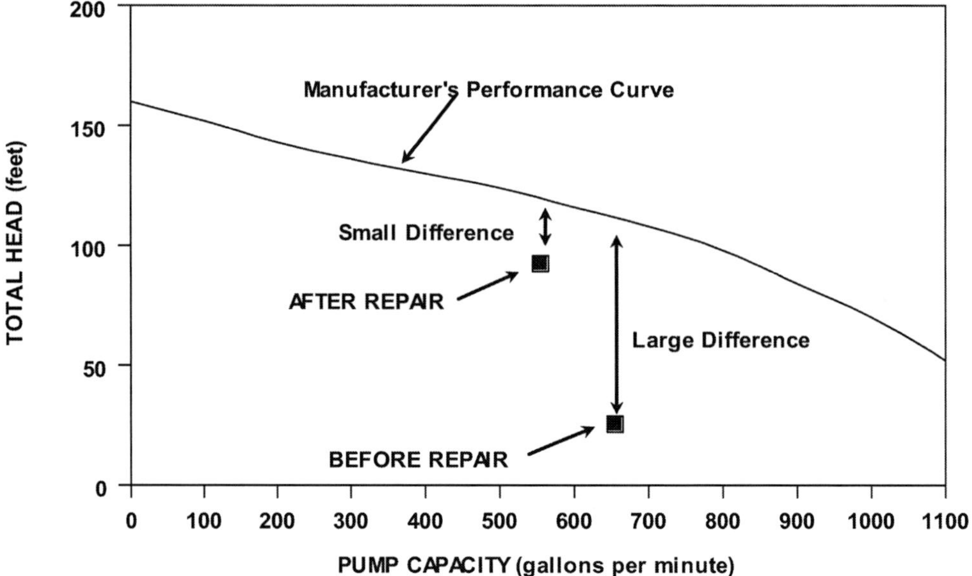

Figure 2. Comparing manufacturer's performance curve to operating points derived from tests on a pump before and after repair.

The following points are important to remember in evaluating a pumping plant:

- A pump can operate over a wide range of conditions.
- A pump test is necessary for determining the status of the pumping plant.
- A pump test requires access to the inside of the well casing and a properly installed flowmeter.
- The pump can be evaluated by comparing the pump test data with the manufacturer's performance curve.

Interpreting Pump Performance Test Results

The following provides guidance on how to use pump performance test data to correct problems.

Pumping Plant Efficiency

Pumping plant efficiency (E) is an indicator of the potential for improvement. The lower the efficiency, the greater the potential for improving pump output or for decreasing input horsepower. Some recommendations:

Table 1. Recommended corrective action to improve pumping plant efficiency at various efficiency levels.

Efficiency Range	*Corrective Action*
Greater than 60%	*No corrective action*
55% to 60%	*Consider adjusting impeller*
50% to 55%	*Consider adjusting impeller first; consider repairing or replacing pump if adjusting impeller has no effect*
Less than 50%	*Consider repairing or replacing pump*

Comparisons of pump test data before and after repair have shown that when pumping plant efficiencies were more than 60%, repairs increased the efficiency by an average of 5 percentage points; (for instance, from 60% to 65%); when efficiencies were between 55% and 60%, the average efficiency increase was about 9 percentage points; when efficiencies were between 40% and 55%, the average efficiency increase was 16 percentage points; when efficiencies were below 40%, the average increase was 21 percentage points.

Note: Inefficiency can be caused by wear, but it can also be caused by a mismatch between the pump and irrigation system characteristics. Repairing a mismatched pump will not improve efficiency. The pump test data should be compared with the manufacturer's performance data to determine whether the pump is mismatched. (See "Pump Performance Tests," page 69, and "Trouble-Shooting," page 79.)

Compare Capacity to How Much Water Is Needed

Even if the pumping plant is operating at high efficiency, it may not provide enough irrigation water at the desired pressures, so the capacity should be compared to the amount of water required for irrigation. The capacity needed for an adequate irrigation can be estimated from the formula given in "What to Consider in Choosing a Pump," page 61. The discharge pressure must be sufficient to maintain the desired sprinkler or emitter discharges and uniformity.

Comparing Multiple-Year Tests

Trends in a pump's performance can be seen by comparing the manufacturer's performance data to pump test data collected over several years. *Figures 1* and *2* compare measured total head with the performance curve values for two pumps over a three-year period. *Figure 1* shows that even though annual efficiency values ranged between 54 and 64%, the difference between measured head and that of the performance curve for Pump A was fairly constant each year, suggesting that the pump is operating properly, with no evidence of declining performance. But *Figure 2* shows an increasing difference between measured head and the performance curve head for Pump B, suggesting that the pump is becoming worn.

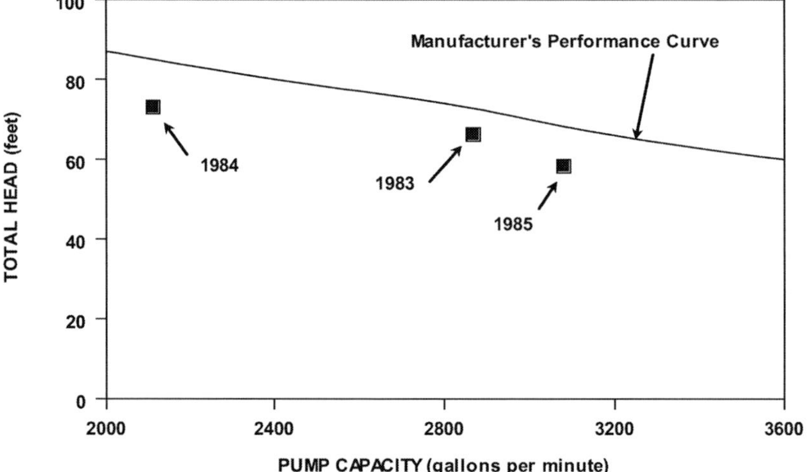

Figure 1. Comparison of manufacturer's performance data to pump test data for Pump A.

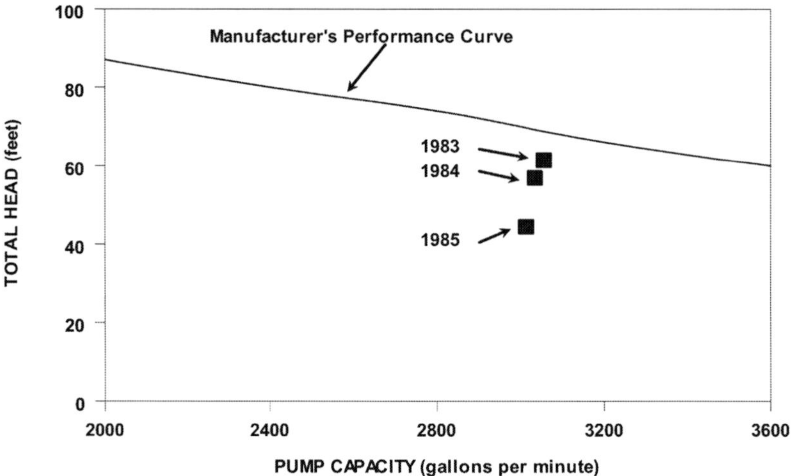

Figure 2. Comparison of manufacturer's performance data to pump test data for Pump B.

Pump and Electric Motor Maintenance

Proper maintenance of your pumping plant will increase the life of the system and ensure adequate performance. Remember, SAFETY FIRST when you are considering any type of pump/motor inspection and/or maintenance. If you don't know how to do something, have a trained and qualified technician either show you how, or do it for you.

Maintaining the Electric Motor

- Periodically check oil levels in motor. The cold oil level should be at the lower mark on the visilube (glass window on side of motor). During the motor operation, the oil level in the visilube will increase because the rotating bearing throws oil away from the bearing into an outer chamber.
- Change motor oil of the upper bearing once per year. Use the proper ASE oil, and use nonfoaming oil
- Do not overgrease the lower motor bushing. To prevent overgreasing, remove the plug, insert a pencil, and inject grease until the pencil moves.
- Avoid overloading the motor. While the service factor indicates that the motor load might be increased a certain amount, any overload will shorten the life of the motor.
- Periodically, check for unusual motor noises, use a "stethoscope-type" engine probe to monitor bearing and machinery "noise" in your pumping plant motor and pump.
- Learn to do an amp check or install amp and volt meters. Changing pumping plant conditions can cause changes in the amperes. Changing or unbalanced voltages will cause varying current loads which can cause motor damage. A volt/phase/current monitor which is programmable and which uses CT's to monitor actual motor current load (such as Motor Saver Model 777) will help monitor and prevent failures due to changing electrical conditions and motor loads.
- On an annual basis, have a qualified electrician check all of the electrical connections, motor starter contacts, etc., including the connections in the motor junction box. This inspection will reveal loose, corroded and defective system wiring and connections. Tightening the connections, replacing corroded and failing components will help increase and lengthen motor life.

Maintaining Your Pump

- Maintain records of your pumping plant. Such record include:
  - Operation and maintenance manual
  - Pump performance curves
  - Motor or engine specifications
  - Pump setting depth

  - Depth of well, well log, casing diameter, type of screen, locations of perforations
  - Pump tests.
- Periodically test the pump performance by conducting a pump test.
- Maintain the appropriate oil drip rate into the column pipe of deep well turbine pumps. A drip rate of 4 to 6 drips per minute per 100 feet of column pipe is recommended.
- Install a flow meter to monitor pump capacity.
- Keep records of pump service and repairs.

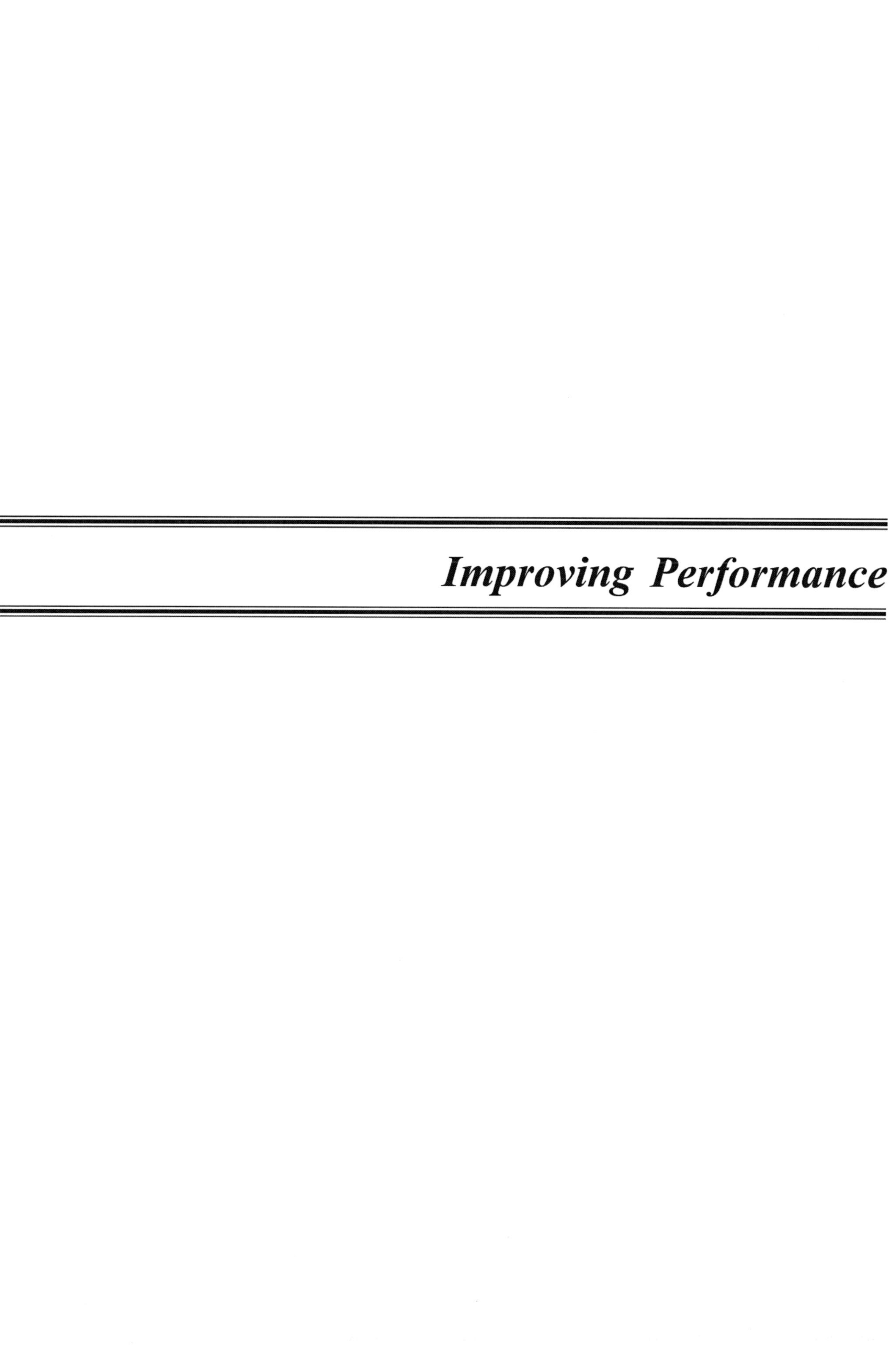

Improving Performance

Trouble-Shooting

Any of the following circumstances should signal to the operator that a pump may not be operating efficiently:

- Sand is being pumped.
- Pump surges or contains air.
- Changes have been made in the irrigation system.
- Water levels in the well have changed.
- No pump test, service, or adjustment has taken place for at least five years.
- A second-hand pump has been installed.
- Several pumps are discharging into a common pipeline.
- Pump discharge is throttled — closed down with a valve — so that full discharge is not being pumped.

Consult the trouble-shooting guide on the following pages for suggested remedies for the most common causes of poor pump performance.

Table 1. Trouble-Shooting Guide

Problem	What to Do
Sand. The presence of sand in the water can greatly accelerate wear of the impellers and pump housing, which in turn cuts down efficiency by reducing pump capacity and total head. Input horsepower may also be slightly decreased. Sand is the principal cause of pump wear. Sand can enter the water as a result of improper well design, poor well construction, or deterioration. As pumping lowers the water level, water flows into the well through small voids in the soil. If water is moving through these voids at high velocity, sand particles can be carried along with the water. A gravel pack is usually installed around the well casing to prevent sand from moving into the well with the water, but if there are large voids in the gravel pack, or if the water is moving with enough force, large amounts of sand can enter the well. Corrosion enlarges the openings in the well casing or screen, allowing sand to migrate into the well.	• *Repair or replace worn pump.* • *Install a sand separator on the pump intake pipe.* Sand separators cause sand to settle out of the water. Sand collects at the bottom of the separator and is periodically discharged to the bottom of the well. Sand separators can be installed in deep-well turbine, line-shaft, submersible, and centrifugal pumps. • *Gradually decrease the pump flowrate before shutting off the pump.* Turning off the pump suddenly when sand is present can cause large amounts of sand, suspended in the water inside the column pipe, to settle out in the impellers and pump housing. Sand can "lock" the impellers, preventing the pump from starting. If this happens, try raising the impellers up and down by using the adjusting nut on top of the motor. Gradually reducing the flowrate before turning off the pump reduces the amount of sand in the column pipe and prevents sand from settling in the impellers and bowls. • *Use semi-open impellers,* which can be raised to adjust the clearance between the impeller and housing. A relatively large clearance can cut down on wear from sand. Note however, that increasing the clearance will reduce total head, capacity, and efficiency, and lower input horsepower. • If sand is entering the well because of a corroded well casing, *repair the well.*
• ***Second-Hand Pump.*** A second-hand pump may not operate at maximum efficiency.	• *Install a pump that provides the desired capacity and total head at maximum efficiency.*

<table>
<tr><th colspan="2">Table 1. Trouble-Shooting Guide (continued)</th></tr>
<tr><th>Problem</th><th>What to Do</th></tr>
<tr><td>

Declining Groundwater Levels. Declining groundwater levels caused by drought, overdraft, or seasonal fluctuations can impair pump performance. As *Figure 1* below illustrates, the greater the pumping lift, the lower the pump capacity. A pumping lift of 33 feet, for example, provides a capacity of 600 gpm, but if the lift increases to 60 feet, pump capacity falls to 200 gpm. In a pressurized system, an increase in pumping lift will cause both pump capacity and discharge pressure to drop.

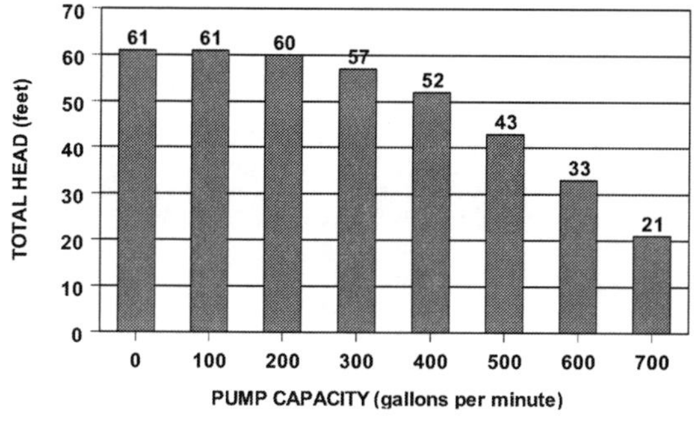

Figure 1. Relationship between pumping lift and capacity.

If the water level falls so low that suction no longer occurs in the intake of the deep well turbine, air may enter the pump intake, causing capacity to fall. When the water level in the well rises, water will again be pumped until the water level falls again and air again enters the intake, resulting in a surging effect. *(See also **Surging,** below.)*

</td><td>

• If the water level falls too low for suction to occur in the intake of the deep well turbine, *lower the pump.* Note that if the pumping level is sufficient to prevent air from entering the pump intake, lowering the pump will have little effect on pump performance. Lowering the pump will cause more head to be lost as a result of friction in the column pipe, which may slightly decrease capacity.

• To keep pump capacity consistent despite variations in pumping levels, *choose a pump with a steep total head/capacity performance curve.*

• *Decrease pump capacity to reduce drawdown in the well* — by installing a valve in the pump discharge, decreasing the engine rpm of an engine-driven pump, raising the semi-open impeller, or by trimming impellers.

• If declining levels are short-term, as during a drought, *increase discharge pressure with a booster pump.* The pressure provided by the booster pump should equal the difference between the existing pressure and the normal operating pressure. A booster pump can also provide some suction lift, which will decrease the pumping lift provided by the deep-well turbine and increase the capacity of the pumping plant. Note however that if the suction lift imposed on the booster pump exceeds 20-25 feet, the performance of the pumping plant will be severely degraded. The total head provided by the booster pump should equal the discharge pressure and the suction lift.

• If declining levels are long-term because of groundwater overdraft, restore pump performance by *removing the pump from the well and installing another pump stage.* Since this measure will increase both the pump capacity and the horsepower demand of the

</td></tr>
</table>

***Table 1. Trouble-Shooting Guide** (continued)*

Problem	***What To Do***
• ***Declining Groundwater Levels*** *(continued).*	pump, a larger electric motor and engine will be required. *Adjust the irrigation time.* Since declining groundwater levels cause pump capacity to decrease, if the irrigation time remains the same, less and less water will be applied as the water level drops. For the same amount of water to be applied, the irrigation time must be increased. Use a flowmeter to determine how much to increase the irrigation time. The new irrigation time can be calculated from the following formula: $t = \frac{(449)(a)(d)}{Q}$ where: t = actual irrigation time in hours Q = pump capacity in gallons per minute a = acres irrigated d = depth to be applied in inches
• ***Surging.*** *(See also **Declining Groundwater Levels,** above.)* Surging occurs when the pump attempts to remove water faster than water is flowing into the well, causing the pumping level to drop to the pump intake. This breaks the intake suction, allowing a slug of air to enter the pump, which decreases the pump capacity and causes the pumping level to rise. Water then enters the pump and pump capacity increases, repeating the cycle and creating the surging effect.	• *Lower the pump capacity to a rate compatible with the flowrate into the well.* • If encrustation has clogged the openings in the well casing and thereby slowed the flowrate, *rehabilitating the well* can improve the flowrate and may eliminate the surging. *(See also **Encrustation,** below)* • If surging is caused by a decline in groundwater level that has lowered the pumping level to that of the pump intake, *lower the pump or reduce the pump capacity to lessen the drawdown.*
• ***Wear.*** Wear can change the performance characteristics of a pump, resulting in less total head for a given pump capacity and cutting down pump or pumping plant efficiency. The principal cause of pump wear is the presence of sand in the water.	• *See **Sand** (above)*

***Table 1. Trouble-Shooting Guide** (continued)*

Problem	***What To Do***
• ***Cascading Water.*** When some of the water-bearing strata lies above the pumping level, water entering the well cascades down to the pumping level. This falling water can distort pump test results. Pumping water levels are usually measured with electric water level indicators, which include electrodes that are lowered into the well. When the electrode contacts the water surface, the electrical circuit is closed, causing the current to flow through an ammeter. If the electrode instead contacts cascading water, the resulting pumping level reading will not be accurate.	• *Use an air line to measure pumping water levels where cascading water may distort pump test results.*
Entrained Air. Cascading water *(see above)* can also cause entrained air — a mixture of small bubbles and water that develops when water falling down to the pumping level traps air. If the water is moving downward at a high enough velocity, bubbles will be carried down into the pump intake. Entrained air reduces pump capacity by causing water to be displaced from the impeller.	• *Decreasing the distance of fall* — either by decreasing the pump capacity to reduce the drawdown or by increasing the number of openings through which water flows into the well — may alleviate problems caused by entrained air. • *Install baffles in the well to break up the falling water.*
• ***Cavitation.*** Cavitation occurs when a vacuum develops in the pump, causing water vapor bubbles to form, which collapse violently when they reach the impeller. The vacuum develops when the rotation of the impeller lowers the water pressure at the impeller eye or center, causing water to flow into the pump intake. If the pump is positioned above the well water level and pressure in the impeller eye is less than that of the atmosphere, a vacuum results. Water vapor bubbles form if the vacuum becomes too large. The noise in severe cavitation resembles the sound of rocks or pebbles rattling inside the pump. Cavitation causes pitting of the impeller and pump housing and can greatly degrade pump performance.	• *Lessen the distance between the pump intake and the water surface.* Limit the suction lift to 20-25 feet. • *Decrease the pump capacity* to lower the flowrate in the impeller. This measure will increase the pressure at the impeller eye and lessen the amount of pressure lost to friction in the intake pipe. • *Increase the diameter of the intake pipe and remove any valves, elbows, and other fixtures.*

Table 1. Trouble-Shooting Guide (continued)

Problem	What to Do
Encrustation. Reactions involving bacteria and minerals in the water can cause minerals to become encrusted in the well casing perforations, in the well screen openings, and in the voids of the gravel pack and water-bearing soil, clogging the openings, impeding water flow into the well, and impairing well efficiency. Encrustation increases the drawdown in the well (the difference between the static and pumping water levels), decreases pump capacity, and increases pumping costs. Iron and manganese hydroxides, which form a jelly-like material, and iron and calcium precipitates are the usual causes of encrustation. The following groundwater characteristics signal encrustation potential: • pH greater than 7.5 • carbonate hardness greater than 300 mg/l or 300 ppm, indicating that calcium carbonate precipitation (white cement-like material) is likely • iron concentration greater than 0.5 mg/ or 0.5 ppm, indicating that iron precipitation (reddish precipitate) is likely • manganese concentration greater than 0.2 mg/l and a high pH, indicating that manganese precipitation is likely	• *Increase the inlet area for flow into the well to reduce the inlet water velocity.* • *Make openings clean-cut and uniform.* • *Chemically treat the well with acids, chlorine compounds and polyphosphates.*[1] [1] For more information on treating wells, *see* V.H. Scott and J.C Scalmanini. *Water Wells and Pumps: Their Design, Construction, Operation, and Maintenance,* University of California Bulletin No. 1889, Division of Agricultural and Natural Resources, University of California, 1978. *See also* "Improving Well Performance," page 93 in this manual.

Table 1. Trouble-Shooting Guide *(continued)*

Problem	***What to Do***
• ***Corrosion.*** Corrosion is the deterioration and destruction of metal screens, casings, and pumps by chemical and galvanic reactions. Chemical corrosion forms a metal compound by dissolving metal, which is then carried away by the water. Galvanic corrosion is caused by electrolytic cells forming between dissimilar metals or other surfaces. The following groundwater characteristics signal corrosion potential: • pH less than 7 • dissolved oxygen content greater than 2 mg/l • hydrogen sulfide in the water (indicated by a rotten-egg odor) • total dissolved solids greater than 1000 mg/l or electrical conductivity of the water exceeding 1.5 dS/m (mmhos/cm) • chloride concentrations exceeding 500 mg/l	• *Use corrosion-resistant materials such as stainless steel or other steel alloys and thermoplastic materials.* • *Prevent contact between dissimilar materials.*
• ***Changes in the Irrigation System.*** Installing additional sprinkler laterals without replacing the pump causes the pump to operate at a higher capacity, at a lower pressure, and probably with less efficiency. The performance of the sprinkler system usually declines as a result.	• *Install a pump with adequate capacity and total head operating at maximum efficiency.* • *Change the irrigation system to match the pump output.*

Table 1. Trouble-Shooting Guide *(continued)*

Problem	***What to Do***
Poor Suction. The amount of inlet suction lift a pump can provide is limited, and if that limit is approached pump performance can decline, causing cavitation in the impellers — a condition in which a large vacuum in the water causes water vapor bubbles to develop. *(See* ***Cavitation,*** *above).* The collapse of these water vapor bubbles can damage the impellers.	• Reduce suction lift by *moving the pump down closer to the water level.* • Cut down the amount of pressure lost as a result of friction in the intake pipe by *removing valves or enlarging the intake pipe.* • *Decrease the flowrate.*
• ***Clogged Impellers.*** Debris or small animals in the water can clog the impellers, greatly reducing pump output and causing vibration, which can damage the pumping plant.	• *Remove blockage.* • *Install a screen over the pump intake* to prevent material from entering the pump.
• ***Insufficient Pressure and Capacity.***	• *Increase speed.* • *Adjust impellers (see page 91).* • Air may be entrained (trapped) in the pump, causing bubbles to be carried down into the pump intake, displacing water from the impeller and reducing pump capacity. *See* ***Entrained Air,*** *above.* • *Check to see whether impellers are trimmed correctly (that is, that the impellers are the correct diameter).* • *Check for leaks in the column pipe.* • *Make certain pump is primed.* • *Check for an air pocket in the suction line.*

Table 1. Trouble-Shooting Guide *(continued)*

Problem	***What to Do***
• ***Excessive Power Demand.***	• In an engine-driven pump, *reduce speed.* • *Check to see that impeller is adjusted properly (See* "Adjusting Impellers," *page 91)* • *Check to see whether pump is out of alignment or if shaft is bent.* • *Check to determine whether sand is being pumped (See* ***Sand,*** *above).* • *Check for imbalance in the electrical motor.*
• ***Vibration/Abnormal Noise.***	• *Realign pump, discharge head, column, and bowls.* • *Check for bent shaft.* • *Check for loose impeller* • *Suspect cavitation — see* ***Cavitation,*** *above.* • *Suspect clogged impellers — see* ***Clogged Impellers,*** *above.*
• ***Pumping Air.*** Air in the pumped water can degrade pump performance by reducing pump capacity. Irrigation system performance can be affected as a result. The main problems associated with air are surging and entrained air.	*See* ***Surging,*** *above.* • *See* ***Entrained Air,*** *above.*

References

Driscoll, F.G. 1986. *Groundwater and Wells.* Johnson Filtration Systems, Inc., St. Paul, Minnesota 55112.

Hydraulic handbook. 1974. Colt Industries, Fairbanks Morse Pump Division.
Irrigation. The Irrigation Association, Silver Spring, Maryland.

Vertical Pump Trouble-shooting Manual. Johnston Pump Company, Glendora, California.

Changing Pump Performance Characteristics

Under certain circumstances — for instance, when a sprinkler system is converted to a lower pressure — it may be desirable to change a pump's performance characteristics (total head and capacity). This can be accomplished by decreasing the impeller diameter (called trimming the impeller) and, in engine-driven pumps, by changing the pump rpm.

Changing Impeller Diameter

If the current pump performance characteristics are known (from manufacturer's curves or pump tests), the effect of a change in diameter on performance characteristics can be estimated using sets of equations known as *affinity laws,* which are given below:

$$Q2 = Q1 \times \frac{D2}{D1}$$

$$H2 = H1 \times \frac{D2^2}{D1^2}$$

$$BHP2 = BHP1 \times \frac{D2^3}{D1^3}$$

where D1 = original impeller diameter
D2 = trimmed impeller diameter
Q1, H1, BHP1 = capacity, head, and brake horsepower of the original performance characteristics (impeller diameter = D1).
Q2, H2, BHP2 = capacity, head, and brake horsepower of pump with trimmed impeller (diameter = D2).

These equations are reasonably accurate if the change in impeller diameter is less than 10%. To use the equations, determine Q1, H1, and BHP1 from the current performance curves or from a pump test and insert these values into the equations to calculate Q2, H2, and BHP2 to develop new performance curves for the smaller impeller.

Example

From pump test data, D1 = 10-7/16" (10.44"); Q1 = 350 gpm, H1 = 105 feet; BHP = 15. Calculate Q2, H2, and BHP2 for an impeller diameter of 9-5/8" (9.62").

$$Q2 = 350 \times \frac{9.62}{10.44} = 322 \text{ gpm}$$

$$H2 = 105 \times \frac{9.62^2}{10.44^2.} = 89 \text{ feet}$$

$$BHP2 = 15 \times \frac{9.62^3}{10.4^3} = 11.7$$

Changing Pump Rpm

The equations for estimating the effect on performance characteristics of changing pump rpm are as follows:

$$Q2 = Q1 \times \frac{RPM2}{RPM1}$$

$$H2 = H1 \times \frac{RPM2^2}{RPM1^2}$$

$$BHP2 = BHP1 \times \frac{RPM2^3}{RPM1^3}$$

where RPM1 and RPM2 = pump rpm before and after change, respectively.

Example

Calculate the total head, capacity, input horsepower, and pumping plant efficiency if the rpm is reduced from 1770 to 1345 with a variable frequency drive. At 1100 gpm the pump develops 185 feet of total head at 1770 rpm. Input horsepower is 128 and overall efficiency is 40%.

$$Q2 = 1100 \times \frac{1345}{1770} = 836 \text{ gpm}$$

$$H2 = 185 \times \frac{1345^2}{1770^2} = 107 \text{ feet}$$

$$IHP2 = 128 \times \frac{1345^3}{1770^3} = 56$$

Adjusting Impellers

Wear can cut down pump efficiency by increasing the clearance between the impeller and the pump housing, thereby decreasing both pump capacity and total head. Wear also damages the surface and shape of the impeller and the vanes in the pump housing.

Adjusting the impellers can partially restore the capacity and total head of pumps having semi-open impellers that are slightly or moderately worn. Adjustment consists of slightly lowering the pump shaft to which the impellers are connected by turning the nut at the top of the motor. The adjustment must also account for lineshaft elongation. This stretching of the lineshaft is caused by the total head, which creates a downward force on the impeller and thus on the line shaft. For pumping plants where total head can vary due to changing pumping and irrigation system conditions, the impeller adjustment should be done under the condition of maximum total head. The cost of an adjustment, which should be performed by a pump dealer, runs between $100 and $200.

Table 1 illustrates the potential effect of adjusting impellers by showing pump test data before and after adjustment.

Table 1. Effect of impeller adjustment on pump capacity, total head, efficiency, and horsepower.

		Capacity (gpm)	Total Head (feet)	Efficiency (%)	Horsepower
Pump 1	Before	605	148	54	42
	After	910	152	71	49
Pump 2	Before	708	181	59	55
	After	789	206	63	65
Pump 3	Before	432	302	54	61
	After	539	323	65	67
Pump 4	Before	616	488	57	133
	After	796	489	68	144

The test data show that capacity and efficiency greatly increased as a result of the adjustment. In two of the pumps, total head also increased. The first pump, an open discharge into a ditch, showed little change in total head.

Adjusting the impeller will not result in an energy cost savings unless the pump operating time of the pump is shortened after the adjustment. In these tests the adjustment increased the input horsepower of each pump. The pumping plant will therefore use slightly more kilowatt-hours — and energy costs will slightly increase — if the operating time is not changed after the adjustment. The main benefit of adjusting the impeller is gaining increased total head and capacity — meaning that more water can be pumped in the same amount of time — which may increase crop yield.

Reference

Sulek, J. J. 1963. "Evaluating factors affecting field efficiency of irrigation turbines sensitive to impeller adjustment." Transactions of the American Society of Agricultural Engineers, Vol. 6(3): 228-223.

Installing Pumps in Series and in Parallel

Installing pumps in series can increase pressure and capacity, and therefore provide greater output, while installing pumps in parallel can increase flowrate.

Pumps are said to be installed in series when one pump discharges into the intake of another pump (*Figure 1*). The flowrate through both pumps is the same, but the total head developed by the second pump is added to the total head of the first pump. The second pump is therefore often called the "booster pump."

One example of pumps in series would be a deep-well turbine pump in series with a centrifugal pump. The turbine pump lifts the water to the surface and the centrifugal pump or booster pump provides the water pressure needed for the irrigation system.

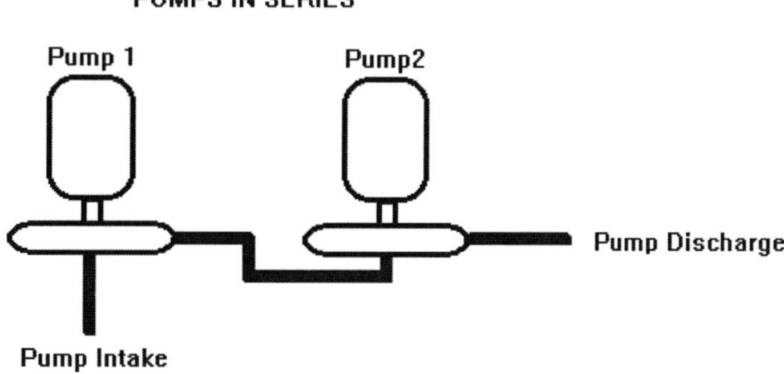

Figure 1. Pumps in series.

The total head/capacity performance curve can be determined by simply adding the total head of each pump at a particular capacity, as shown in *Figure 2*. The booster pump selected should be one that will operate at maximum efficiency at the desired flowrate.

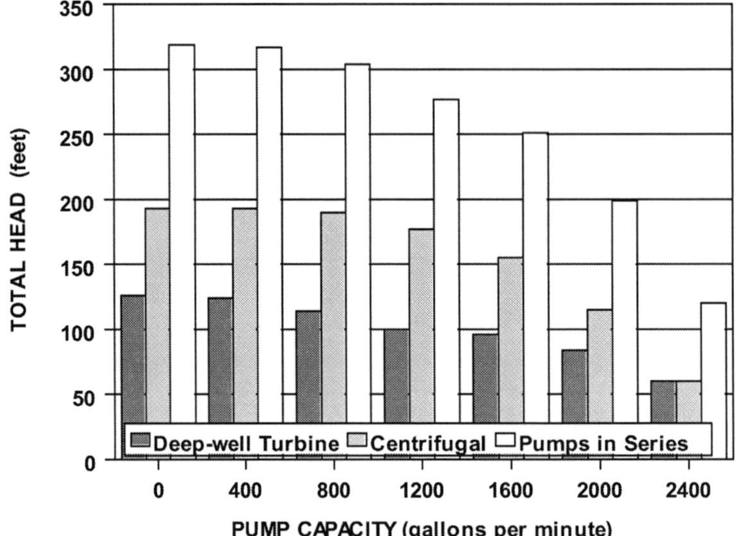

Figure 2. Total head and capacity of pumps in series.

Pumps are said to be *in parallel* when several pumps are discharging into a common pipeline (*Figure 3*). The total head developed by each pump is about the same. The flowrate in the pipeline is the sum of the flowrates of each pump. If two pumps pumping into a pipeline have the same performance characteristics, the flowrate through each pump will be about half the total flowrate. If the pumps are dissimilar, their capacities will differ.

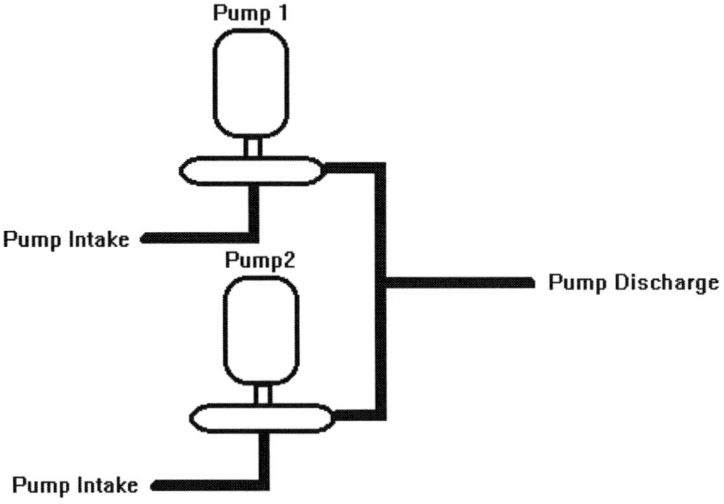

Figure 3. Pumps in parallel.

The total head/capacity performance curve of a pumping plant with pumps in parallel can be determined by adding the flowrate of each pump at a particular total head. *Figure 4*, showing two pumps with the same characteristics, illustrates this. If, however, the performance characteristics of the pumps are not the same, it is possible for total head produced by one pump to be much larger than that of the other pump, causing one pump to override the other. When this happens, the pump being overridden will have no output, although an input horsepower demand will still exist. The overridden pump will simply operate at shutoff.

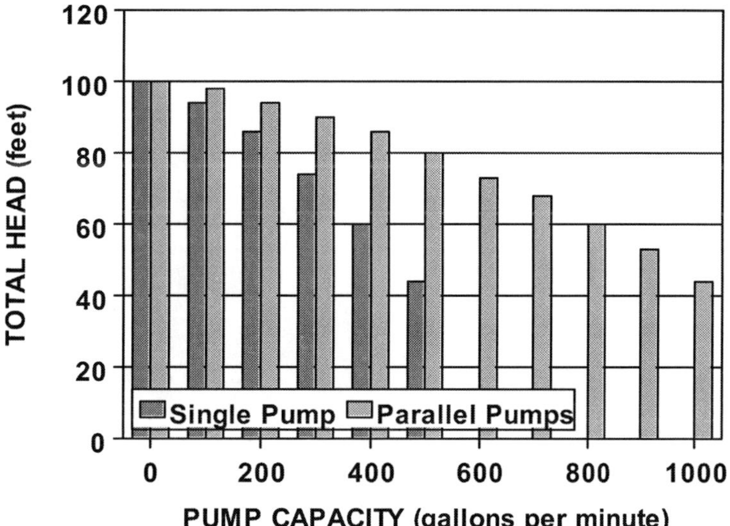

Figure 4. Total head and capacity of a single pump and parallel pumps.

Which Approach is Best?

Sometimes growers expand sprinkler irrigation systems by adding more laterals, resulting in a lower flowrate per lateral, which in turn makes the volume of water applied inadequate. The question then arises: should a second pump be installed in series or in parallel to increase the flowrate?

In this situation, installing a pump in series will provide a higher output than installing a pump in parallel. Getting more water through the sprinkler system requires much more pressure to overcome pressure losses from friction in the pipelines and in the sprinkler nozzle. Installing a pump in series can greatly increase the pressure, which in turn will substantially increase the flowrate. In contrast, installing a pump in parallel will increase the pressure only slightly and will cause a slight increase in flow, but a substantial increase in horsepower demand.

Interactions Between Pumping Plant and Irrigation System

Pump performance curves show the relationships between pump capacity and total head, efficiency, and brake horsepower. However, the actual operating point of the pump will also depend on the performance characteristics of the irrigation system. The irrigation system performance characteristics depends on pumping lift, elevation differences throughout the field, pressure requirements of sprinklers, drip emitters, etc., and friction losses in pipelines and fittings. The performance characteristics of an irrigation system are described by an irrigation system head curve which shows the total head needed by the irrigation system for various system capacities.

After a pump is turned on, the pump output of total head and capacity eventually will come into equilibrium with the irrigation system demand. At that time, the output of the pump equals the capacity and head demand of the irrigation system. The intersection point of the pump total head curve and the irrigation system head curve shows the pump output.

Developing a system head curve for an irrigation system can be a time-consuming procedure unless one has access to computer programs for irrigation system design. Thus, for many irrigation pumping plant operators, system head curves probably will not be available. However, it may be useful to understand some of the interactions that might occur between pumping plants and irrigation systems to help understand the possible effects of changes in the irrigation system or pump on total head and capacity. Several examples illustrate some of these interactions.

System Head Curve

Figure 1 shows the total head curve for a pump and the system head curve for a sprinkler system. As discussed in earlier chapters, pump total head decreases as capacity increases. However, for the sprinkler system, total head increases as capacity increases, simply because more head is needed to discharge more water out of the sprinklers and to overcome increased frictions losses in the pipeline and fittings. The pumping lift reflects the head needed to get the water to the surface assuming that groundwater is being used for irrigation.

At equilibrium, the total head and capacity are 95 feet and 1360 gallons per minute, respectively. The intersection point of the two curves is at Point A (*Figure 1*). About 77 feet of total head is available to pressurize the irrigation system.

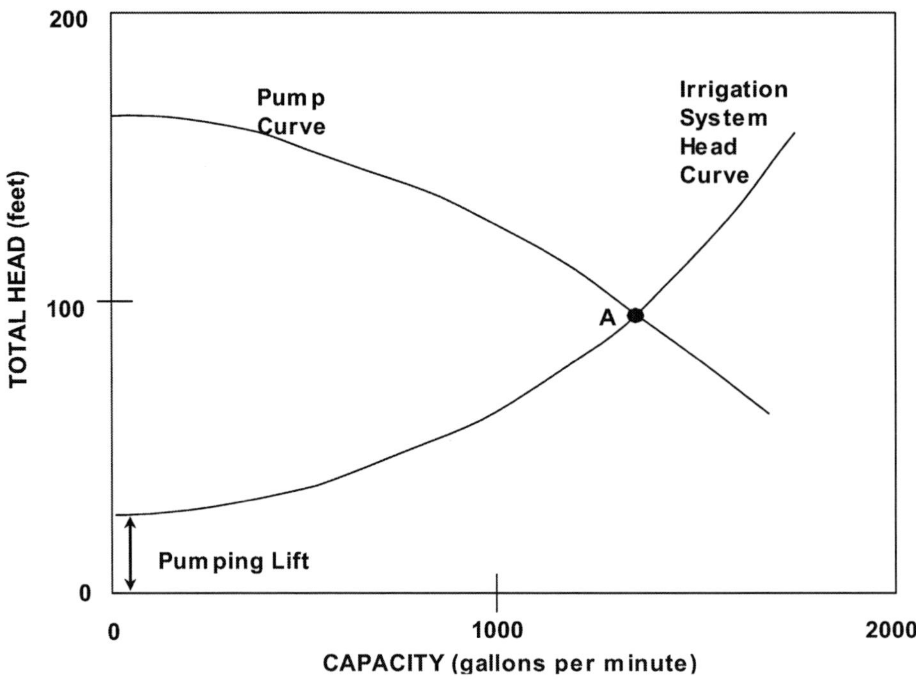

Figure 1. Pump total head and irrigation system head curves.

Declining Goundwater Level

Figure 2 shows the effect of a declining groundwater level on the performance of pump and irrigation system. Because more head is now needed to lift the water to the ground surface, the irrigation system head curve has been shifted upward by an amount equal to the difference between the old and new pumping lifts. The equilibrium point is at Point B where the total head is 104 feet and the capacity is about 1200 gallons per minute. However, only about 50 feet of head is are available for system pressurization.

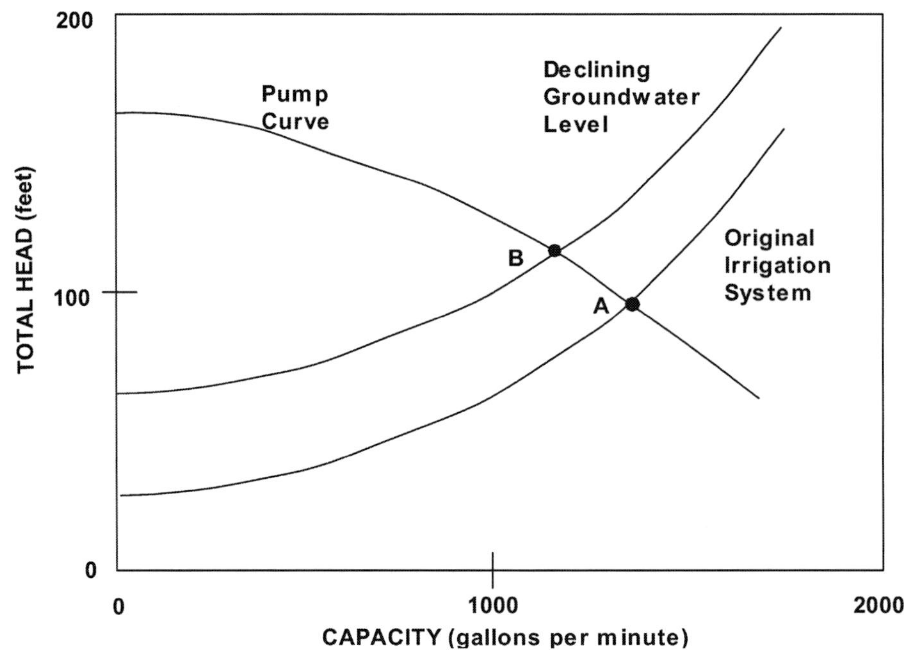

Figure 2. Performance under declining groundwater conditions.

Expanded Irrigation System

Sometimes, the irrigation system is expanded beyond the original design, but the same pumping plant continues to be used. *Figure 3* shows pump and irrigation curves for an expanded sprinkler irrigation system. The new equilibrium point is now Point C. At this point, system capacity is about 1530 gallons per minute and total head is about 77 feet. However, the head available for system pressure is only 55 feet. It can be seen that the expansion caused the pump's capacity to increase, but at the same time total head decreased because of the pump's performance characteristic.

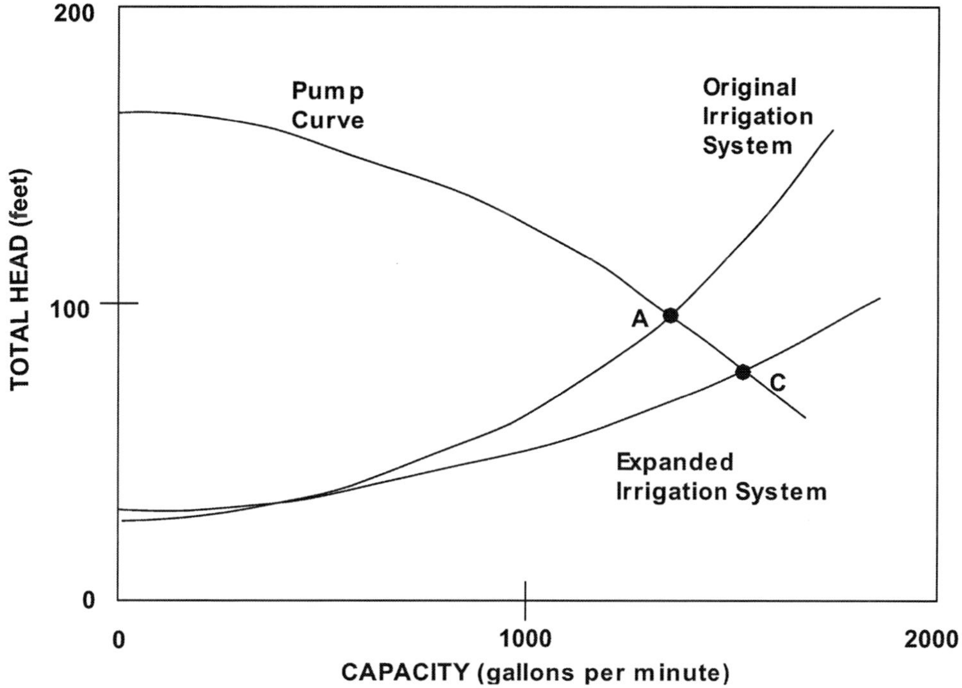

Figure 3. Effect of changes in irrigation system.

Pumps in Series

Frequently, a centrifugal pump is placed in line or in series with a deep well turbine pump to increase pressure. The effect of this practice is illustrated in *Figure 4*. Individual pump head curves are shown for the two pumps along with the head curve of the two pumps in series. The equilibrium point of the pumping plant with the irrigation system is at Point D, where the total head and capacity are 127 feet and 1600 gallons per minute, respectively.

Pumps in Parallel

In some cases, pumping plants used to pump irrigation water for canals and ponds have been modified to increase the capacity by adding a second pump placed in parallel with the original pump. The effect of this practice is illustrated in *Figure 5* which shows the pump total head curve of a single pump and the total head curve of two pump in parallel. It is assumed that both pumps are the same model and have the same output. The equilibrium point for the parallel system is at Point E, where total head and capacity are 136 feet and 1670 gallons per minute, respectively, compared to 95 feet and 1360 gallons per minute of the original pumping plant. However, the extra 310 gallons per minute is expensive water because the horsepower requirement of the pumping plant is now doubled. One might be better off in the long run buying a new pump capable of supplying the output of the two pumps in parallel.

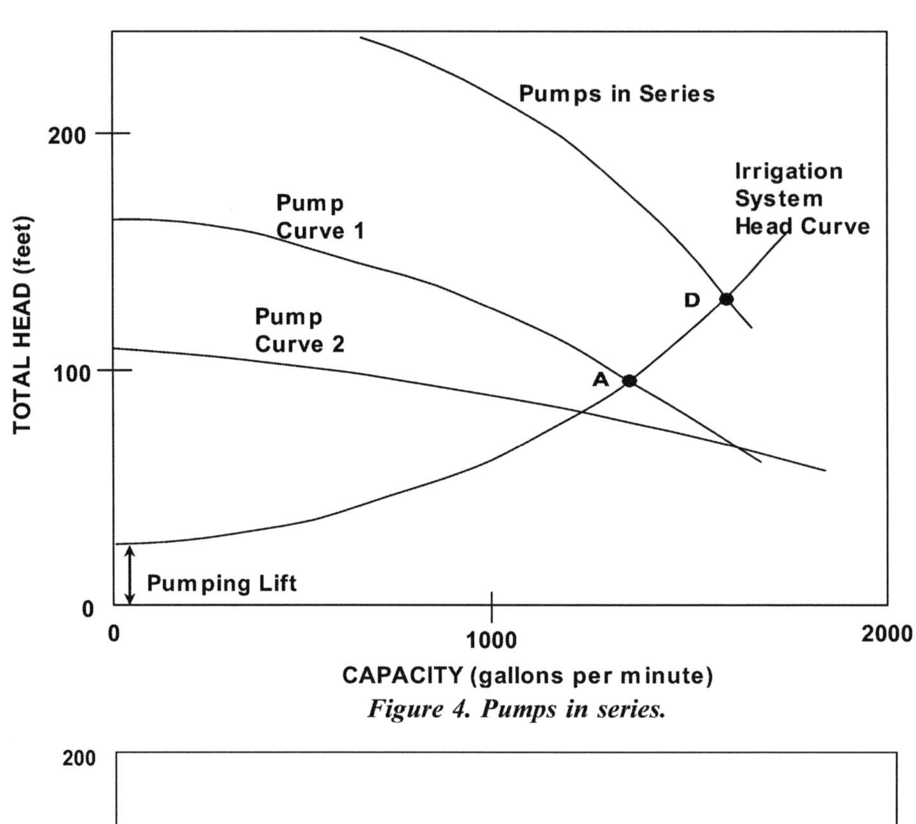

Figure 4. Pumps in series.

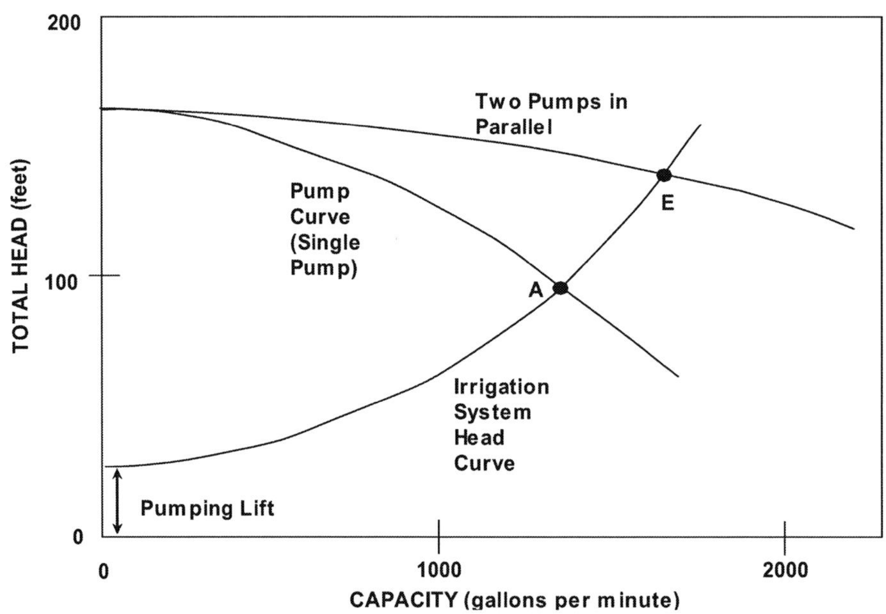

Figure 5. Pumps in parallel.

These examples show qualitative responses to changes in irrigation system or pumping plants to help irrigation pumping plant operators understand potential effects of the changes. Magnitudes of the changes will differ for other pumps and irrigation systems.

Improving Well Performance

The efficiency of a well is defined by the pumping water level compared to what the pumping level would be if the water flow into the well was unrestricted. High efficiency means less drawdown at a given pump capacity, lower energy costs, or more pressure and capacity at a given pump horsepower. Actually, well efficiency estimates are rarely performed because the process is difficult and expensive.

The following design factors influence the relative efficiency of a new well:

• ***Depth.*** The well should penetrate the water-bearing material sufficiently to provide the desired capacity with minimum drawdown — the difference between the static or non-pumping water level and the pumping water level.

• ***Open area available for flow into the well.*** The more open area in the well casing available for flow into the well, the greater the efficiency. Well screens provide the most open area. An open area of about 15 percent is acceptable. Well openings may also be louvered or slotted (mill-sawed, torch-cut, punched, or perforated in place). Slotted openings usually have poor uniformity and are not efficient.

• ***Location of openings in relation to the water-bearing formations.***

• ***Gravel pack design.*** The gravel pack used should retain water-bearing material without causing excessive hydraulic loss. Procedures for selecting a gravel pack are given in the first of the two references cited below.

• ***Well development.*** The well must be developed in such a way as to remove any drilling fluids from the water-bearing material and to stabilize the gravel pack and aquifer material around the well. Development methods include bailing, mechanical surging, air surging, and jetting. One study of nine wells showed that surging a newly constructed well increased the specific capacity of the well (gallons per minute per foot of drawdown) by about 23 percent. A combination of surging and jetting increased the specific capacity by nearly 37 percent.

The efficiency of older wells hinges on good maintenance, with encrustation and corrosion the principal factors diminishing performance. Encrustation, which results from the precipitation of carbonates, from iron and manganese precipitates, and from reactions involving bacteria in the water, can decrease the open area in the well casing, plugs well screens, and clog the voids in the gravel pack.

Determining the Cause of Encrustation

The type of deposits evident in the well are a guide to the cause of the encrustation:

• Red or brown deposits are caused by iron in the presence of bacteria.

• Black deposits are caused by iron and sulfide or by manganese in the water.

• White or yellow deposits are caused by calcium carbonate or magnesium carbonate.

Treating Encrustation

The following rehabilitative measures can be effective in treating wells affected by encrustation.

• ***Use chlorine*** to disinfect wells affected by bacterial growth and slimes. Calcium hypochlorite and sodium hypochlorite are the materials commonly used for chlorination. Disinfection must take place before acid treatment is applied.

• ***Inject acid*** into the well to dissolve the encrusting material. Inhibited hydrochloric acid and sulfamic acid are commonly used.

• ***Use polyphosphates*** to disperse clay and silt particles that might be clogging the gravel pack or water-bearing material.

• ***Use mechanical development techniques*** — brushing, surging, jetting, and air lift pumping — after chemical treatments have been applied.

Treating Corrosion

Corrosion is the deterioration of the metal casing or screen, which can result in sand entering the well. Sulfate-reducing bacteria can cause corrosion in wells that remain unused for long periods of time. Correcting the problem may require replacing the damaged components.

Effect of Well Performance on Energy Costs

Poor well efficiency can affect pump performance and raise energy costs. *Table 1* shows data from pumping plant tests conducted before and after a pump repair. The worn pump was installed in an old well. While the pump was being removed the well casing collapsed. A new well was drilled and the repaired pump was installed in the new well. The overall efficiency of the worn pump was 39 percent, while that of the repaired pump was 50 percent.

Table 1. Pumping plant test data before and after pump repair.

	Before	*After*
Pumping lift (feet)	*45*	*102*
Pressure (psi)	*1*	*1*
Capacity (gpm)	*624*	*537*
Input horsepower	*19*	*28*
Overall efficiency (%)	*39*	*50*
Energy use (kwhr/AF	*123*	*211*

Although the repair substantially improved the pump performance in this example, energy use following the repair increased by almost 71 percent — from 123 kwhr per acre-foot of pumped water before the repair to 211 kwhr per acre-foot after the repair. The main reason for this energy increase was that the well was poorly constructed, as evidenced by the test data. In the new well, the pumping lift was 102 feet at a pump capacity of 537 gallons per minute, while in the old well, the pumping lift was 45 feet at a higher capacity. The high energy costs therefore resulted from the poor efficiency of the new well.

References

Scott, V.H., and J. C. Scalmanini. 1978. *Water wells and pumps: Their design, construction, operation, and maintenance.* University of California Bulletin No. 1889, Division of Agricultural Sciences. University of California.

Werner, H.D., T.F. Scherer, and T.O. Kajer. 1980. "Irrigation well design and development to improve efficiency." American Society of Agricultural Engineers Paper 80-2093, presented at the ASAE 1980 Summer Meeting, San Antonio, Texas, 15-18 June 1980.

Department of Land, Air and Water Resources, University of California, Davis

Case Studies: Pumping Plant Evaluations

The following describes evaluations of actual pumping plants and suggests improvements.

Worn Pump

Situation: A two-stage deep-well turbine pump was tested under open discharge conditions. The overall efficiency was 36 percent. Test results were as follows:

Table 1. Pumping plant test data on a two-stage deep-well turbine under open discharge conditions.

Pumping lift (feet)	*51*
Capacity (gpm)	*2134*
Input horsepower	*77.4*
Overall efficiency (%)	*36*

Evaluation and Corrective Action: The low efficiency may have been caused by a worn or mismatched pump or other factors. An initial assessment of the problem, made by comparing the test data with the manufacturer's performance data, is reflected in *Figure 1.* The analysis reveals a significant difference between the performance curve (about 115 feet of total head at 2134 gpm) and the test data (51 feet of total head at 2134 gpm), suggesting that the pump is worn and should be repaired or replaced. A lesser difference would suggest the pump is mismatched to the capacity and head.

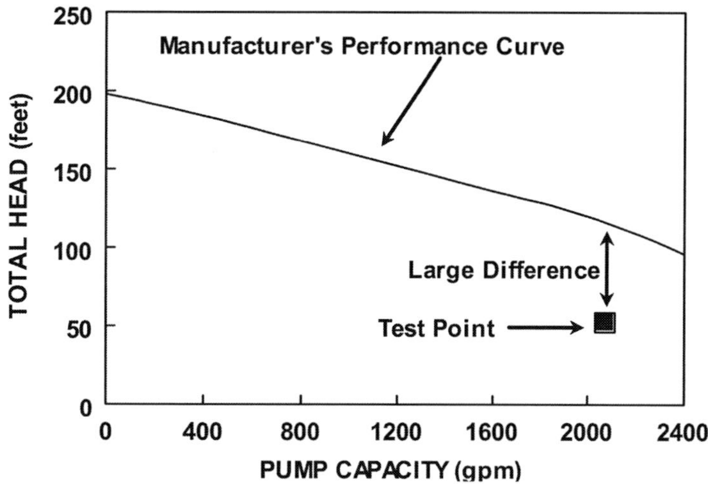

Figure 1. Comparison of pump test data and manufacturer's performance data.

Pump Mismatched to Operating Conditions

Situation: A second-hand pump was installed in a well to supply water to a sprinkler irrigation system. Both capacity and pressure were adequate for the irrigation system, but pump test data showed the overall efficiency to be 48 percent. Test data were as follows:

Table 2. Pumping plant test data for a second-hand pump installed in a well to supply water to a sprinkler irrigation system.

Pumping lift (feet)	*113*
Discharge pressure (psi)	*50*
Total head (feet)	*228*
Capacity (gpm)	*940*
Input horsepower	*111.9*
Overall efficiency (%)	*48*

Evaluation and Corrective Action: The low efficiency may have stemmed from improper adjustment, mismatch to conditions, worn pump, or other factors. Since the manufacturer of this pump was not known, no performance curves were available. To overcome this problem, several pump tests were carried out at differing capacities. At 870 gpm, the overall efficiency was an acceptable 57 percent. This higher efficiency suggests that the pump may have been operating properly, but was mismatched to the pumping conditions — a common problem when a second-hand pump is used.

Corrective action is to replace the pump with one that can supply 940 gpm at 228 feet of total head at a higher efficiency. If the pump selected had had an efficiency rate of 57 percent, the input horsepower would have been 95 instead of 112. This would have saved more than 20,000 kilowatt-hours per year for this grower. Energy costs would have been reduced by $1008 per year at $0.05/kwhr., or $2016 per year at $0.10 kwhr. The payback period would have been about one year.

Clogged Impeller

Situation: An engine-driven centrifugal pump supplied water from a ditch to a gated pipe. The flowrate dropped from 2200 gpm to 1500 gpm during the irrigation.

Evaluation and Corrective Action: This situation suggests a clogged impeller. Inspection of the impeller revealed a frog caught in the vanes. An intake screen should be installed to protect the pump from foreign objects.

Excessive Power Costs After Repairing Pump

Situation: A two-stage deep-well turbine was found to have an overall efficiency of 39 percent, which is poor. The pump was repaired, but during the removal of the pump the well casing collapsed. A new well was drilled nearby and the repaired pump was installed. The overall efficiency of the repaired pump was 50 percent, but the energy consumption of the repaired pump was almost twice that of the worn pump. Test data were as follows:

Table 3. Pumping plant test data comparing a worn and a repaired pump, showing excessive energy consumption in the repaired pump.

	Worn Pump	*Repaired Pump*
Pumping lift (feet)	*45*	*102*
Discharge pressure (psi)	*1.5*	*1.4*
Capacity (gpm)	*624*	*537*
Input horsepower	*18.9*	*28*
Overall efficiency (%)	*39*	*50*
Energy use (kwhr /AF)	*123*	*211*
(AF = acre-feet)		

Evaluation and Corrective Action: Figure 2 compares the manufacturer's performance curve with test data before and after repair. The difference between the curve and test data of the worn pump suggests that the low efficiency was caused by pump wear. The difference between the curve and test data on the repaired pump indicates that the pump is operating properly.

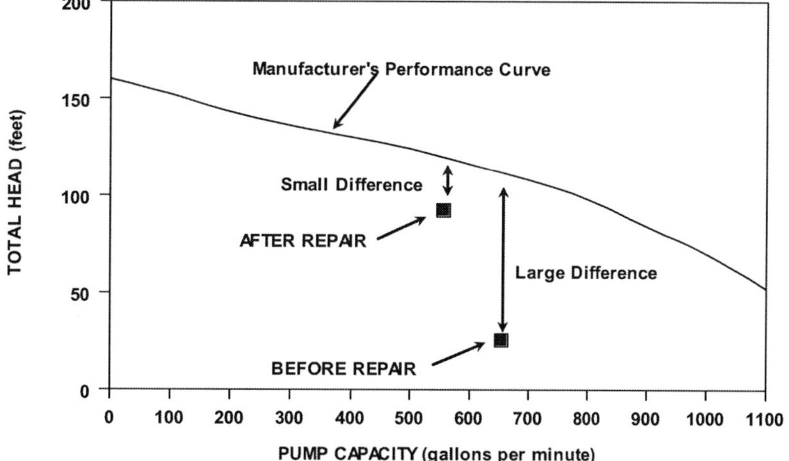

Figure 2. Comparison of manufacturer's performance curve with test data before and after repair.

The test results show that the pumping lift in the old well was 45 feet at 624 gpm, but in the new well the pumping lift was 102 feet, even though the pump capacity was lower, indicating that more of the pump output was being used for pumping lift compared to the old pump. This large increase in pumping lift was probably caused by an improperly constructed well. The flowrate was also reduced and the input horsepower was substantially increased — from about 19 to 28. This combination of factors resulted in a much higher energy use per acre-foot of water pumped.

Little can be done to correct the poor performance of this well. Proper construction would have avoided the problem.

Inadequate Pipe Test Section

Situation: Tests on a centrifugal pump pumping out of a slough into a sprinkler irrigation system found overall efficiency to be about 70 percent. Energy use was calculated to be about 216 kwhr per acre-foot. The pump test report noted that the absence of a sufficiently long straight section of pipe made conditions poor for measuring flowrate. The grower reported, however, that actual energy use was much higher than that reported in the test. A second pump test was conducted using sprinkler nozzle discharge measurements to estimate the flowrate. Test results were as follows:

Table 4. Pumping plant test data on a centrifugal pump where an inadequate pipe test section resulted in an inaccurate flowrate measurement.

	First Test	*Second Test*
Suction lift (feet)	5.5	5.5
Discharge pressure (psi)	62	79
Total head (feet)	149	188
Capacity (gpm)	547	227
Input horsepower	22	29
Overall efficiency (%)	70	37
Energy use (kwhr /AF)	216	500

Evaluation and Corrective Action: A comparison of test data and performance curve showed that in the first test, the operating point was considerably above the performance curve and beyond what the pump was capable of producing. This reflects the error in the flowrate measurements caused by the poor test conditions. The second test shows an operating point below the performance curve. There may be an error in this flow measurement, which was obtained from sprinkler discharges, but it was nonetheless more realistic than that obtained from the first flowrate measurement, since the energy use of 500 kwhr/AF reflected the grower's actual costs.

Cascading Water

Situation: A new pump installed in a well was found through testing to have a low efficiency of 33 percent. The test report noted cascading water in the well — a condition in which water falls into the well from a point above the water level. The pump was removed, repaired, and reinstalled. A second test revealed the same efficiency level. Both tests used an electric water level indicator to measure the pumping level. A third pump test was conducted using an air line to measure the pumping level. Test results were as follows:

Table 5. Cascading water in a well.

	First Test	*Third Test*
Pumping level (feet)	42	81
Discharge pressure (psi)	0.5	5
Total head (feet)	43	92
Capacity (gpm)	1414	1178
Input horsepower	47	47
Overall efficiency (%)	33	58

Evaluation and Corrective Action: The cascading water prevented accurate measurement of the pumping lift during the first and second tests because the electrodes on the electric water level indicator used for the measurements contacted the cascading water in the well at about 43 feet, interpreting that water as the water at the pumping level. Air line measurements, though, unaffected by the cascading water, reported a pumping lift of 81 feet, nearly twice that of the first test. The poor efficiency reflected in the first test resulted from inaccurate measurements rather than from pump performance.

Change in Sprinkler System Nozzle Size Decreases Overall Efficiency

Situation: Sprinkler nozzles were replaced with nozzles of a larger diameter. Pump tests conducted before and after the nozzle change showed a decrease in the overall efficiency, which was of major concern to the grower. Test data were as follows:

Table 6. Pumping plant test data showing a decrease in overall efficiency after sprinkler nozzles were replaced with nozzles of a larger diameter.

	Before	*After*
Pumping lift (feet)	93	98
Discharge pressure (psi)	41	25
Total head (feet)	187	157
Capacity (gpm)	383	440
Input horsepower	29	34
Overall efficiency (%)	61	51
Energy use (kwhr/AF	313	317

Evaluation and Corrective Action: As expected, the nozzle change caused the discharge pressure to decrease and the pump capacity to increase. Input horsepower also increased, reflecting the brake horsepower performance curve. Overall efficiency decreased by 10 percentage points.

Although the overall efficiency decreased, energy use per acre-foot of pumped water remained about the same because of the increased capacity and the increased input horsepower. However, if annual operating hours remain the same after the change, the increased input horsepower will cause the annual energy cost to increase.

The principal benefit of this change is the increased water supply. If the output of the "after" pumping plant is adequate, then changing to a more efficient pump (61 percent efficiency) will reduce the input horsepower from 34 to 28.6. The energy saved at operating 1600 hours per year will be about 7100 kwhr — about $710 per year at $0.10 per kilowatt-hour.

Pump Capacity Inadequate for Expanded Irrigation System

Situation: Water was supplied to a sprinkler irrigation system by a centrifugal pump pumping from a slough. Pump test data are given below. The irrigation system had been expanded from four laterals to six laterals. As a result, flowrate per lateral decreased and crop growth was poor. The grower wanted to install a second pump (same model and type) in parallel with the first to increase the flowrate. The length of the mainline between the pump and the first lateral was about 2100 feet.

Table 7. Pumping plant test data on a centrifugal pump pumping from a slough into a sprinkler irrigation system where adding laterals made pump capacity inadequate.

Discharge pressure (psi)	*69*
Suction lift (feet)	*7*
Total head (feet)	*166*
Capacity (gpm)	*1386*
Input horsepower	*87*
Overall efficiency (%)	*67*

Evaluation and Corrective Action: The pump test data show that the pump is operating properly; pump performance cannot be improved. An analysis of two pumps in series and two pumps in parallel using the manufacturer's performance curve showed that a second pump in series with the first could provide about 1840 gpm, whereas a second pump in parallel could provide about 1600 gpm. A pump in series, though, would provide too much pressure unless the pump were located near the first lateral. In either case, the increase in pumping plant output will be expensive, since adding a pump in series or in parallel will double the horsepower demand.

Benefits of Improving Pumping Plant Efficiency

The main benefit of improving pumping plant efficiency is increased pump output — resulting in greater pump capacity and, in some cases, greater total head. This will result in fewer kilowatt-hours expended per acre-foot of pumped water. (Total head will increase with pressurized irrigation systems, but may remain unchanged with low-pressure irrigation systems.) Test data on sixty-three pumps in the Salinas Valley revealed that repair or replacement increased capacity by an average of 41 percent. Total head increased only slightly, mainly because of the low-head irrigation systems (gated pipe) used in that region. Pumping plant efficiency increased by an average of 33 percent and input horsepower increased by about 8 percent. *Table 1* illustrates the effect of pump repair on one pumping plant.

Table 1. Effect of pump repair (bowls rebuilt) on pump performance.

	Before	*After*
Pumping lift (feet)	*170*	*179*
Pumping capacity (gpm)	*506*	*933*
Input horsepower	*70*	*71*
Overall efficiency (%)	*31*	*59*

Will Improving Pumping Plant Efficiency Save Energy?

Energy charges are figured on kilowatt-hours used, which consist of input horsepower (or kilowatts) and the operating time of the pump in hours. If energy is to be saved, either kilowatts or horsepower and/or operating time must be reduced. If the improvement causes little change in the input horsepower and the operating hours remain the same, kilowatt-hours will also remain unchanged despite the increase in pumping plant efficiency, and no energy savings will result.

• ***Retaining the operating time.*** As *Table 1* illustrates, improving the pumping plant efficiency may have little effect on the input horsepower. Kilowatt demand will therefore be the same before and after the repair. If the same operating time is in effect before and after the repair, the number of kilowatt-hours consumed will be unchanged, regardless of the improvement in pump efficiency, and no energy will be saved. The benefit of improving pumping plant efficiency under these conditions will be higher capacity and greater total head, which may improve crop yield.

• ***Reducing the operating time.*** Because of the increased capacity, the operating time can be reduced while allowing the same volume of water to be pumped. As a result, kilowatt-hours will decrease and energy will be saved. Options for reducing operating time are:

• *Decrease the irrigation set time.* Many growers are reluctant to decrease set times because of labor problems arising from the need to change sets during the night and from the need to keep track of set times of other than eight, twelve, or twenty-four hours.

• *Increase the acreage irrigated per set, taking advantage of the increased pump capacity and reducing the irrigation time for the field.* (This may not be feasible for pressurized irrigation systems.) Energy will be saved only if the number of irrigation sets is decreased.

• ***Replacing the pump.*** Replacing a worn pump with a new, efficient pump that provides the same output will lower energy costs by decreasing horsepower or kilowatt demand, so long as the output (particularly the pump capacity) is adequate for the irrigation system. Using the pump described in *Table 1* as an example, if the pump output before the repair is sufficient, then substituting a new pump that provides the same output at an overall efficiency of 65% will lower the input horsepower to 33, compared to 71 horsepower for the repaired pump.

Points to Remember

Energy will be saved only if the kilowatt-hours used by the pumping plant are reduced.

• If the pump is mismatched to the operating conditions (is not operating at close to maximum efficiency), replacing it with a properly matched pump will reduce the input horsepower and result in energy savings even if the operating time remains the same.

• Repairing or replacing a worn pump can substantially increase pump capacity, total head, and efficiency in pressurized irrigation systems. In low-pressure systems, including those using gated pipe, total head may change only slightly. However, input horsepower may change only slightly as a result of the repair. Thus, kilowatt-hours will be reduced only by decreasing the pumping time.

Time-Of-Use Electric Rates

Time-of-use (TOU) electric rates are designed to encourage pumping during off-peak electric use periods and to discourage pumping during periods of peak use. The incentives offered are low electric rates for off-peak pumping and penalties in the form of very high electric rates and high demand charges for on-peak pumping.

The peak-use period is usually set at between noon and 6 pm on weekdays, while the off-peak period is normally between 6 pm and noon of the following day, although how peak and off-peak hours are defined depends on the utility company. Some TOU rate schedules include a partial-peak period between the on- and off-peak hours. Winter and summer rates may also differ.

Operating at a TOU rate for 18 hours a day instead of at a flat rate for 24 hours a day can result in a considerable energy cost savings. Whether converting to an 18-hour-per-day operation is feasible, however, depends on the pump capacity and the irrigation system. To avoid applying less water per set and thereby possibly affecting crop yield, the pumping plant must have enough capacity to apply the same volume of water in 18 hours that would be applied in 24 hours. Sprinkler and micro-irrigation systems may have to be modified to increase application rates for 18-hour-per-day operation.

The complexity of rate schedules and rate variation among utility companies make presenting a detailed description of TOU rates here impractical. For more information about TOU rates and to evaluate the suitability of TOU rates in a given situation, utility companies should be contacted directly.

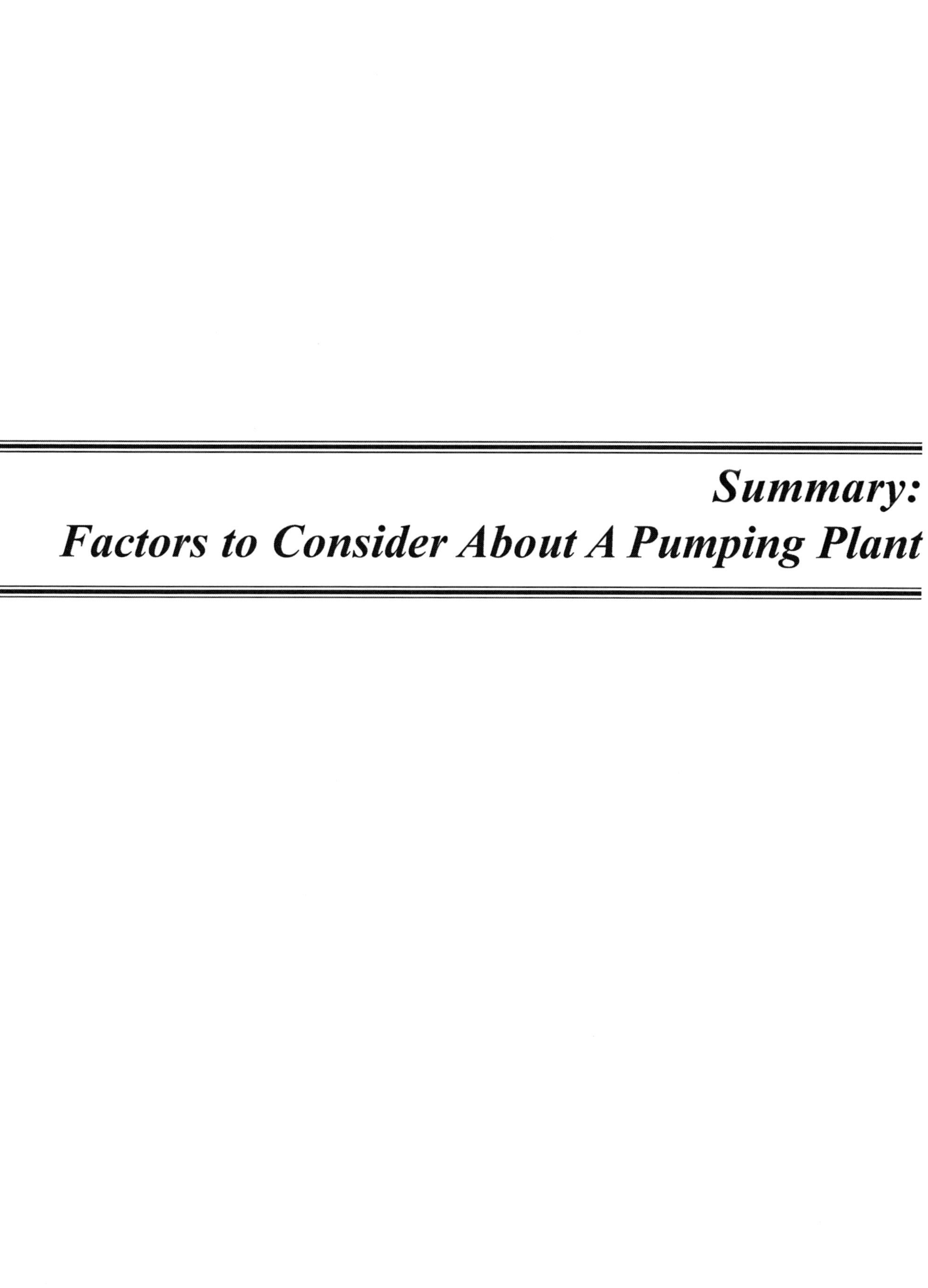

Summary: Factors to Consider About A Pumping Plant

Summary: Factors to Consider About A Pumping Plant

- ***New Installation:***
    - Will the pump operate at close to maximum efficiency at the desired head and capacity?
    - Will the pump capacity be adequate for my irrigation needs?
    - Will the pressure be adequate for my irrigation needs?
    - Will the motor or engine horsepower be adequate?
    - Will the engine operate at close to maximum efficiency?
- ***Existing Installation:***
    - What is the efficiency of the pumping plant?
    - Is the capacity and pressure of the pump adequate?
    - If pumping plant efficiency is low, what is the cause?

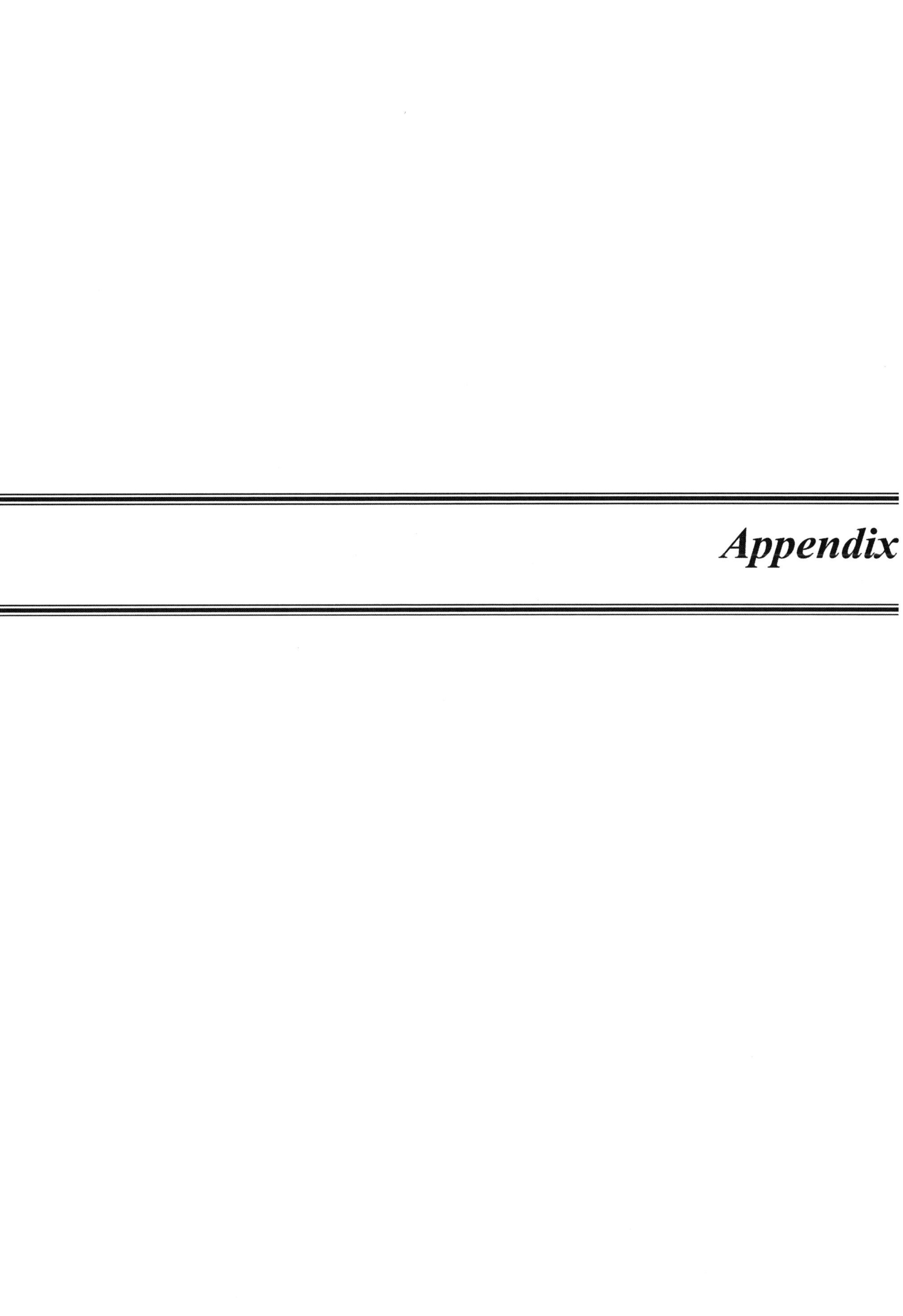

Appendix

Bibliography

Driscoll, Fletcher G., Editor, *Groundwater and Wells.* 1989. Johnson Filtration Systems, Inc., St. Paul, Minnesota 55112.

Groundwater Manual. 1977. U.S. Government Printing Office, Washington, D.C. 20402, stock number 024-003-00106-6.

Halderman, D.A. 1979. *Submersible vs. Line-shaft Vertical Irrigation Pumps.* University of Arizona Leaflet 79-1. University of Arizona, Tucson.

Hansen, H.J. and W. L. Trimmer. 1986. *Extending Electric Motor Life.* Pacific Northwest Extension Publication No. 292.

Hydraulic Handbook. 1974. Colt Industries, Fairbanks Morse Pump Division.

Irrigation. The Irrigation Association, Silver Spring, Maryland.

Irrigation Pumping Plant and Well Efficiency Handbook. 1982. Department of Agricultural Engineering, University of Nebraska, Lincoln, Nebraska 68583-0726. (402) 472-2824.

Knutsen. G., R. Curley, and W. Chancellor. 1981. *Selecting Electric Motors for Maximum Efficiency.* University of California Leaflet 21240.

Knutsen, Jerry. 1981. "Converting From Electrical Power to Diesel Power for Irrigation." Department of Agricultural Engineering, University of California, Davis.

Pacific Gas and Electric Company. 1985. *Three-Phase Service to Submersible Pumps.* San Francisco, California.

Scott, V.H., and J.C. Scalmanini. 1978. *Water Wells and Pumps: Their Design, Construction, Operation, and Maintenance.* 1978. University of California Bulletin No. 1889, Division of Agricultural Sciences, University of California.

Vertical Pump Trouble-Shooting Manual. Johnston Pump Company, Glendora, California.

Werner, H.D., T. F. Scherer, and T.O. Kajer. 1980. "Irrigation Well Design and Development to Improve Efficiency." American Society of Agricultural Engineers Paper 80-2093, presented at the ASAE 1980 Summer Meeting, San Antonio, Texas, 15-18 June 1980.

Glossary

Acidizing. Injecting acid — usually hydrochloric acid or sulfamic acid— into a well to dissolve encrusting material.

Air Surging. *See* ***Surging.***

Axial flow pump. Pump consisting of pump housing, impellers, and shaft. Axial flow pumps, sometimes called *propeller pumps,* are used for low-head, high capacity conditions.

Balance line. Often included in centrifugal pumps to help counterbalance unequal forces on the impeller.

Cascading water. Water entering the well at a point above the water level.

Cavitation. The formation of a vacuum that occurs when a pump is positioned above the well water level and pressure in the impeller eye is less than that of the atmosphere. Cavitation results if the vacuum becomes too large, causing bubbles of water vapor to be formed, which collapse violently, making a loud rumbling noise, when they reach the discharge side of the impeller. Cavitation causes pitting of the impeller and pump housing and can greatly degrade pump performance.

Centrifugal pump. Also called a "booster pump." Centrifugal pumps are placed at the ground surface and are often used for pressurizing irrigation systems and for pumping from surface water sources. They consist of the volute (impeller housing), impeller, and shaft.

Chlorination. Treating wells with chlorine compounds such as calcium hypochlorite or sodium hypochlorite to counteract the growth of bacteria and slimes.

Corrosion. Deterioration and destruction of metal screens, casings, and pumps by chemical and galvanic reactions. Chemical corrosion forms a metal compound by dissolving the metal, which is then carried away by the water. Chemical corrosion can allow sand to enter the well. Galvanic corrosion is caused by electrolytic cells forming between dissimilar metals or surfaces.

Deep-well turbine. Pumps installed inside the well casing and submerged below the water level in the well. Deep-well turbines consist of bowls, impellers (located inside the bowls), shaft (connecting the impellers to the power source), and discharge head. They are often used to pump groundwater.

Discharge pressure. *See* ***Pump discharge pressure.***

Double volute. Volute used in some centrifugal pumps, having a single vane splitting the water flow to help neutralize uneven radial forces acting on the impeller, shaft, and bearings. *See also* ***Volute.***

Drawdown. The difference between static water level and pumping water level.

Efficiency. *See* ***Overall efficiency; Pump efficiency; Pumping plant efficiency; Well Efficiency.***

Encrustation. The accumulation of material in the perforations of the well casing, in well screen openings, and in the voids of the gravel pack and water-bearing soil. Encrustation clogs the openings, decreasing the open area in the well casing, impeding water flow into the well, and decreasing well efficiency.

Entrained air. A mixture of small bubbles and water that develops when some of the water-bearing strata lies above the pumping level. Water entering the well falls down to the pumping level, trapping air. If the downward velocity of the water is sufficient, bubbles will be carried downward into the pump intake. Entrained air displaces water from the impeller, reducing pump capacity.

Friction loss. Loss of pressure as a result of friction caused by water moving through the system.

Head. *See* ***Total head.***

Impeller. Pumping plant component made of bronze, cast iron, plastic, or cast iron coated with porcelain enamel. The impeller transfers energy developed by the motor or engine to the water.

Overall efficiency. The combined efficiency of the pump efficiency and the efficiency of the motor or engine. Pump capacity, multiplied by total head, divided by input horsepower. *(Same as* ***pumping plant efficiency.)***

Performance curve. Measurement available from the pump manufacturer showing the relationship between total head and flowrate. The amount of head developed by a pump at a particular capacity or flowrate.

Performance characteristics. The relationships between a pump's total head and capacity, efficiency and capacity, and brake horsepower and capacity.

Pump capacity. The flowrate in gallons per minute.

Pump discharge pressure. Pressure required by the sprinklers or emitters in addition to the pressure needed to overcome elevation differences and pressure losses caused by friction.

Pump discharge pressure head. Discharge pressure multiplied by 2.31.

Pump efficiency. Efficiency of the pump by itself, determined from manufacturers' performance curves.

Pumping plant efficiency. The combined efficiency of the pump efficiency and the efficiency of the motor or engine. Pump capacity, multiplied by total head, divided by input horsepower. *(Same as **overall efficiency.**)*

Pumping lift. The distance from the discharge pipe at the pump head to the water level in the pumping well.

Pumps in parallel. Several pumps discharging into a common pipeline to increase the flowrate.

Pumps in series. Pumps installed so that one pump discharges into the intake of another pump. The total head developed by the second pump (often called the "booster pump") is added to the total head of the first pump, thereby increasing pressure.

Sand separator. Device installed on the pump intake pipe in deep-well turbine pumps to cause sand to settle out of the water.

Semi-open impeller. Impeller with only one plate or shroud supporting the vanes.

Suction lift. Distance from the water surface to the pump intake when the pump is located above the surface.

Submersible pump. Deep-well turbine with a waterproof electric motor, which is installed in the well below the pumping level.

Surging. Cycle created when a pump attempts to remove water faster than water is flowing into the well, which causes the pumping level to drop to the pump intake, breaking the intake suction and allowing a slug of air to enter the pump. The pump capacity falls and the pumping level begins to rise. Water then enters the pump and the pump capacity increases, causing the cycle to repeat and creating a surging action.

Time-of-use rates. Lower rates for electric power used during off-peak hours and very high rates and high demand charges for power used during peak use hours to encourage pumping during off-peak electric use periods and discourage pumping during periods of peak use.

Total head. The capability of a pumping plant to lift water to the soil surface, combined with the pressure needed to supply water throughout the irrigation system and to overcome pressure losses caused by friction and elevation differences in the system. The pumping lift added to the discharge pressure head.

Volute. Spiral-shaped impeller housing in a centrifugal pump.

Wearing rings. Bronze rings often attached to the pump housing on either side of the impeller in centrifugal pumps.

Well efficiency. The pumping water level in a well compared to what the pumping level would be if water flow into the well was unrestricted.

Index

A

B

C

G

H

I

L

M

N

O

P

R

S

T

U

V

W

Y